EVIDENCE -BASED HEALTH CARE MANAGEMENT
Multivariate Modeling Approaches

EVIDENCE-BASED HEALTH CARE MANAGEMENT
Multivariate Modeling Approaches

by

Thomas T.H. Wan, Ph.D.
Virginia Commonwealth University
USA

KLUWER ACADEMIC PUBLISHERS
Boston / Dordrecht / London

Distributors for North, Central and South America:
Kluwer Academic Publishers
101 Philip Drive
Assinippi Park
Norwell, Massachusetts 02061 USA
Telephone (781) 871-6600
Fax (781) 871-6528
E-Mail <kluwer@wkap.com>

Distributors for all other countries:
Kluwer Academic Publishers Group
Distribution Centre
Post Office Box 322
3300 AH Dordrecht, THE NETHERLANDS
Telephone 31 78 6392 392
Fax 31 78 6546 474
E-Mail <orderdept@wkap.nl>

RA427.9
.W36
2002

047965137

 Electronic Services <http://www.wkap.nl>

Library of Congress Cataloging-in-Publication Data

A C.I.P. Catalogue record for this book is available from the Library of Congress.

Printed on acid-free paper.

Printed in the United States of America

To *Sylvia J. Wan,*
my Wife and Friend

To *William K. Wan* and *George J. Wan,*
my Sons
and
Emmy Wan,
my Daughter-in-Law - -

With much love.

and

In Memory of Clara Chassell Cooper and Sol Levine,
my Mentors - -

A very special thanks.

CONTENTS

List of Tables ix
List of Figures.. xi
About the Author xiii
Foreword xv
Acknowledgments xvii

Chapter 1
An Introduction to Evidence-Based Management of Health Care 1
Chapter 2
Causal Inference: Foundations for Health Care Managerial Decision-
Making 9
Chapter 3
Research on Health Services Management: The Search for Structure .33
Chapter 4
Exploratory Analytical Modeling Methods 47
Chapter 5
Introduction to Structural Equation Modeling 71
Chapter 6
Confirmatory Factor Analysis 89
Chapter 7
Structural Equations for Directly Observed Variables: Recursive
and Non-Recursive Models 119
Chapter 8
Structural Equation Models with Latent Variables 137
Chapter 9
Covariance Structure Models 155
Chapter 10
Multiple Group Comparison with Panel Data 177
Chapter 11
Multilevel Covariance Modeling 201
Chapter 12
Growth Curve Modeling with Longitudinal Data 213
Chapter 13
Epilogue 229

Index 231

LIST OF TABLES

Table 1. Statistical Methods for Analyzing Program Performance, by Study Design and Type of Outcome..20

Table 2. Type I and Type II Errors in Hypothesis Testing....................................24

Table 3. Multiple Logistic Regression Analysis of Mortality Risk for Abdominal Aortic Aneurysm (N = 174 Patients Treated): Two Models 53

Table 4. Three Factor Dimensions and Zero-Order Correlations between Each of the 18 General Well-being (GWB) Items and the Total GWB Scores...59

Table 5. The Greek Alphabet 77

Table 6. The Net Effects of Predictor on Adverse Patient Outcomes 86

Table 7. Goodness of Fit Measures for Proposed and Revised Models of Depression Measurement 104

Table 8. Demographic Characteristics of the Study Sample 111

Table 9. Correlation Matrix of Eight Aggregate Indices.....................................111

Table 10. Parameter Estimates for the Initial and Revised Models.......................112

Table 11. Goodness of Fit Statistics..114

Table 12. Operational Definition of the Variables 128

Table 13. Correlation Coefficients of the Study Variables 130

Table 14. The Decomposition of Effects: Proposed Structural Model..................132

Table 15. The Decomposition of Effects: Revised Structural Model....................134

Table 16. List of the Study Variables and Definitions 145

Table 17. A Measurement Model of Hospital Performance Indicators..................149

Table 18. Structural Equation Model of Hospital Performance: Parameter Estimates 152

Table 19. Standardized Parameter Estimates for the Causal Model of Physician Productivity in a Community Hospital...158

Table 20. Operational Definition & Measurement Instruments for the Study Variables 160

Table 21. A Measurement Model of Two Exogenous Latent Variables (Factor Loadings) 164

Table 22. A Measurement Model of Endogenous Variables (Factor Loadings)...164

Table 23. A Structural Equation Model of the Effects of Exogenous Variables on Caregivers' Burden ($\eta 1$) 164

Table 24. A Measurement Model of Exogenous Latent Variables (with a Buffering Effect)..165

Table 25. A Structural Equation Model With a Buffering Effect of Exogenous Variables on Caregivers' Burden (η_1)...165

Table 26. Goodness of Fit (GOF) Indices 165

Table 27. Definitions of the Study Variables 169

Table 28. The Relationship between Informatic Integration and IDS Inefficiency (N=973)..172

Table 29. Hypothesis Testing of Equality Constraints..177

Table 30. Indicators of Patient Satisfaction in the Ambulatory Care Patient Population 180

Table 31. Definitions of the Study Variables 187
Table 32. Descriptive Statistics and Correlation Matrix of All Nursing Units......190
(N = 124)..190
Table 33. Descriptive Statistics and Correlation Matrix of Nursing Units from
Small Hospitals 190
(n = 62)   190
Table 34. Descriptive Statistics and Correlation Matrix of Nursing Units from
Large Hospitals (n=62) 190
Table 35. Summary Statistics for the Measurement Models of Patient Satisfaction,
with and without Correlated Measurement Error................................192
Table 36. Goodness-of-fit Statistics of Stacked Model...194
Table 37. Summary Statistics for the Generic Structural Equation Model 195
and Revised Model...195
Table 38. Latent Constructs and Their Indicators 205
Table 39. Parameter Estimates for the Two-Level Analysis of Nine Indicators ...209
of Patient Satisfaction..209

LIST OF FIGURES

Figure 1. The Dynamic Nature of Knowledge Management 2
Figure 2. Foundation of Scientific Inquiry...10
Figure 3. The Seven Steps of Scientific Causal Inquiry....................................14
Figure 4. Causal Ordering with Regard to Disease 25
Figure 5. Mutual Causation, or Reciprocal Causal Paths....................................26
Figure 6. Framework for Evaluating Health Care System Performance 36
Figure 7. A Data Warehouse for Health Systems Analysis..................................37
Figure 8. Health care Information for Management...38
Figure 9. Iterative Process in Generating New Knowledge 40
Figure 10. Determinants of Health Services Use 49
Figure 11. A Path Model..61
Figure 12. Path Analytic Model 1 63
Figure 13. Path Analytic Model 3 66
Figure 14. Path Analytic Model 3 67
Figure 15. A Measurement Model 74
Figure 16. A Structural Model 74
Figure 17. Model 1: Physical Functioning 75
Figure 18. Model 2: Mental Functioning 75
Figure 19. Health Services Research Models...75
Figure 20. A Measurement Model of Patient Adverse Outcomes..........................76
Figure 21. A Measurement Model of Health Status...79
Figure 22. A General LISREL Model...81
Figure 23. A Measurement Model of Adverse Health Care.................................83
            Outcomes in Acute Care Hospitals 83
Figure 24. Measurement Model of Health Status/Outcomes 91
Figure 25. A Congeneric Measurement Model 91
Figure 26. Two Sets of Congeneric Models...93
Figure 27. Three Sets of Congeneric Measures...93
Figure 28. Confirmatory Factor Analysis of Health Status................................94
Figure 29. The Generic Measurement Model of Depression 97
Figure 30. A Proposed Measurement Model..103
Figure 31. Revised Measurement Model 105
Figure 32. Model Without Correlated Measurement Errors.................................113
Figure 33. Model With Correlated Measurement Errors......................................113
Figure 34. A Recursive Model 121
Figure 35. Two Unrelated Endogenous Variables With Common Causes 122
Figure 36. A Path Model with Y1 Affecting Y2..122
Figure 37. A Path Model with A Reciprocal Causation 123
Figure 38. Correlated Residuals...123
Figure 39. A Path Model with Correlated Residuals..124
Figure 40. A Recursive Path Model 125
Figure 41. A Multi-Wave Panel Model with Two Utilization Variables 126
Figure 42. Proposed Structural Model 129
Figure 43. The Revised Model...133

Figure 44. A MIMIC Model..138
Figure 45. A Recursive Model of Health Services Use with Social Support.......139
Self Report Functional Status Factors as Predictors............................139
Figure 46. Model with Correlated Measurement Errors...................................140
Figure 47. Autoregression...142
Figure 48. A Nonrecursive Model with Latent Constructs for Exogenous...........143
and Endogenous Variables..143
Figure 49. A Proposed Structural Equation Model of Hospital Performance.......146
Figure 50. Revised Structural Equation Model of Hospital Performance............151
Figure 51. A Covariance Structure Model...156
Figure 52. A Complex Model with Correlated Residuals...............................156
Figure 53. A Proposed Causal Model of Physician Productivity in a Community
Hospital...157
Figure 54. An Epidemiological Model of the Oral-Facial Pain Syndrome..........159
Figure 56. A Measurement Model of the Oral-Facial Pain Syndrome.................161
Figure 57. Structural Equation Model of the Oral-Facial Syndrome.................162
Figure 58. A Covariance Structure Model of Caregiving Burden......................163
Figure 59. A Covariance Structure Model of Caregiving Burden with...............166
an Interaction Term (Buffering Effect)..166
Figure 60. A Covariance Structure Model of the Relationship Between Informatic
Integration and Hospital Performance..170
Figure 61. The Relationship between Informatic Integration and IDS Efficiency 171
Figure 62. A Measurement Model of Physical Well-Being with Panel Data........178
Figure 63. Two-Wave Panel Study of Patient Satisfaction with Quality of Care .181
Figure 64. Two-Wave Panel Study of Patient Satisfaction with Access of Care.181
Figure 65. Two Dimensions of Patient Satisfaction: A Panel Study...................182
Figure 66. A Generic Model for Explaining Poor Physical Health......................183
Figure 67. A Measurement Model of Patient Satisfaction...............................186
Figure 68. Covariance Structure Model of Patient Satisfaction.........................188
Figure 69. Revised Measurement Model of Patient Satisfaction........................191
Figure 70. Measurement Model of Patient Satisfaction with Multiple................193
Group Analysis (Small vs. Large Hospital Size)...................................193
Figure 71. Covariance Structure Model of Patient Satisfaction with Correlated
Errors...194
Figure 72. Revised Covariance Structural Model of Patient Satisfaction............196
Figure 73. Revised Covariance Structural Model of Patient Satisfaction............197
Figure 73. Measurement Model of Patient Satisfaction at the Nursing Unit Level207
Figure 74. A Linear Growth Curve Model of Hospital Occupancy Rates...........219
(1996-1999): Model 1..219
Figure 75. A Linear Growth Curve Model of Hospital Occupancy Rates (1996-
1999) with Time-Invariant and Time Varying Predictors: Model 2....221
Figure 76. A Parallel Process Growth Model for Two Continuous Outcome......224
Variables with Time-Invariant Predictors (1996-1999): Model 3.......224
Figure 77. An Interface between Structural Equation Modeling and...................230
Constraint-Oriented Reasoning Methodology, for Health Care
Knowledge Management and Research...230

ABOUT THE AUTHOR

Thomas T.H. Wan, Ph.D., is Professor, Department of Health Administration, Medical College of Virginia Campus, Virginia Commonwealth University, Richmond, Virginia and held the Arthur Graham Glasgow Chair from 1991 to 1999.

Professor Wan received a Bachelor of Arts in Sociology from Tunghai University, Taiwan; a Master of Arts and a doctorate in Sociology from the University of Georgia, and a Master of Health Sciences from the Johns Hopkins University School of Hygiene and Public Health, where he was also a National Institutes of Health postdoctoral fellow. His over 25 years in academia encompass faculty positions at Cornell University; the University of Maryland, Baltimore County; and his current position at Virginia Commonwealth University, where he has been teaching since 1981.

Also at the University, Professor Wan has served as the founding Director of the Doctoral Program in Health Services Organization and Research, and as Director of the Williamson Institute for Health Studies. He also served as the Associate Editor of the journal *Research on Aging;* a member of the Editorial Board of the *Journal of Gerontology;* a member of the Executive Committee for the Association for Social Scientists in Health; a member of the Governing Council, Medical Care Section, American Public Health Association; a Senior Research Fellow of the National Center for Health Services Research; a member of the Study Section on Aging and Human Development II, National Institute on Aging; a member of the National Committee on Vital and Health Statistics of the Centers for Disease Control and Prevention; and a member of the advisory board of the Veterans Integrated Service Network VI.

Professor Wan is a member of the Study Section on Nursing Research, NIH. He is also a member of the National Health Research Institutes' Scientific Review Committee in Taiwan and an advisor for four IT & informatics companies-- Attotek, Perimed Compliance Corporation, Strategic Medical Alliance Informatics, and MedVersant. He has established collaborative research and educational programs in countries such as Taiwan, China, Korea, Czech Republic, South Africa, Kazakhstan, etc.

His research interests are centered in managerial epidemiology, health services evaluation, health informatics, and clinical outcome studies. His published work includes: 1) *Promoting the Well-being of the Elderly: A Community Diagnosis* (Haworth Press, 1982); 2) *Stressful Life Events and Gerontological Health* (Lexington Books, 1982); 3) *Well-being of the Elderly: Preventive Strategies* (Lexington Books, 1985); and 4) *Analysis and Evaluation of Health Care Systems: An Integrated Managerial Decision Making Approach* (Health Professions Press, 1995).

FOREWORD

Slowly but steadily, the health care system is transitioning into one in which good evidence both is available and is used to stimulate effective performance by health care providers and organizations. Management by tradition, guesswork, imitation, and intuition is being supplemented or supplanted by management based on evidence. Extended government and private foundation support for health services research, and the institutionalization of health services research as a profession, reflect and reinforce this remarkable and necessary transformation of the health care system. The transformation is fueled by the realization that the status quo in health care management is not acceptable. Evidence of patient safety problems observed throughout the health care system, of variation in quality of care, and in the United States, of health indicators that lag those of many other countries, all have contributed to the rise of health services research. Health services research has become an important tool in the promotion of change in the way that health care is managed.

Clinical structures, processes and outcomes have been the target first exposed to the wide variety of scientific methods to prove efficacy and effectiveness. Management structures, processes and outcomes are not far behind. In *Evidence-Based Health Care Management*, Professor Thomas Wan pushes forward on that frontier. A social scientist by education and experience, throughout his long and distinguished career Professor Wan has focused on the use of social science to solve intractable social problems and to advance human welfare. He firmly believes that the world of practice can benefit from the best in social science research design and methods.

As the body of evidence and the feasibility of collecting evidence in health care management advance, so must the methods for examining that evidence and establishing better or best practices. In this book, Professor Wan challenges those studying health care management problems to use the best research methods available. Professor Wan clearly communicates a wide range of sophisticated techniques for multivariate analysis that can more validly tackle the complex management problems faced by health care practitioners and executives.

Professor Wan emphasizes the concept of causality, and unraveling the causal links among interdependent and dependent variables is the goal of the research designs and methods he promotes. At heart, the interest in pursuing causality links managers and researchers arm in arm. Managers "guesstimate" or assume causality in their decisions every day. For simple problems, causal relationships often are clear. As problems intensify in complexity, managers' ability to "guesstimate" falters. Similarly, causality is more difficult for researchers to assess as the complexity of the problem examined increases, calling for more intricate research designs and more sophisticated multivariate tools for analysis of data.

Professor Wan's contribution to the expansion of the foundations of health services management and research is to be applauded *Evidence-Based Health Care Management* should be read with care and openness to the grand potential for science and practice to be more intimately linked for the enhancement of health and health care delivery.

James W. Begun, Ph.D.
Professor and Chair
Director, MHA Program
Department of Healthcare
Management
University of Minnesota

ACKNOWLEDGMENTS

This book will introduce scientific principles and methods for generating evidence-based knowledge for health services management and research. The ideas in this book might not have been formulated without the students who shared their insights and interests to learn advanced quantitative methods. This book is dedicated to health care managers and researchers who have tried or will try to produce scientific evidence to guide the development and implementation of effective and efficient health services. Appreciation must be expressed to my associates--Kwangsoo Lee, Allen Ma, Willis Gee, Marie Gerardo, and Mindy Wyttenback--for their assistance in preparing this book, and to Dorothy Silvers and Beverly DeShazo for their editorial assistance. Finally, I want to thank my wife, Sylvia Wan, for her patience and support in completing this book.

CHAPTER 1

AN INTRODUCTION TO EVIDENCE-BASED MANAGEMENT OF HEALTH CARE

In recent years there has been an explosion of evidence-based medicine/practice. It has been the direct result of several factors: the aging of the population, the increase in patient expectations and professional expectations, the proliferation of new information technologies, and the growth of disease management modeling. Massive amounts of clinical and administrative data have been gathered and have been managed in relational databases to generate information for improving health care processes and outcomes. That information is a systematic repertory of knowledge, used often by providers, administrators, patients, and others (see Figure 1). Used properly and consistently, such evidence-based knowledge, will give users a competitive edge in the health care marketplace.

Evidence-based medicine integrates the best evidence from clinical databases with clinical expertise, pathophysiological knowledge, and patient preferences, in deciding on methods of individual treatments (Ellrodt et al., 1997). That approach has been shown to improve disease management. When evidence is the key factor shaping optimal practice, a more coordinated, holistic system of health care delivery is likely to emerge. Moreover, evidence-based disease management can incorporate the most effective methods of changing clinician behavior (Shortell, et al., 2001). Thus, by reducing the variation in medical practice, evidence-based medicine may improve clinical as well as economic outcomes (Sackett et al., 1996).

While the use of evidence has made its way into the practice of medicine, its introduction into the management of health care has been slow. Muir Gray has stated: "The practice of evidence-based health care enables health services managers to determine the mix of services and procedures that will give the greatest benefit to the population served" (Gray, 1997). The continuing evolution of health care will call on managers to maintain quality yet control costs, while also meeting the demands of a changing population. Technology will drive this evolution. This chapter examines the role of data warehouses in evidence-based management. It seeks to tie together the use of massive information storage centers with the wealth of information—i.e. evidence—not now being used.

In evidence-based decision-making, an important focus is the concept of knowledge management. Though its definition varies from organization to organization, knowledge management is concerned with networks that intertwine people and information. The generation of knowledge can be viewed as a two-part cycle. In the personal cycle, knowledge is acquired by compiling information from reports, memos, and papers. Information derived from these sources is then filtered through the individual's experience and expertise. In the collective cycle, information is not only acquired, but also shared and acted on. The age of technology permits collection of massive data; unfortunately, however, much of it remains unused. Yet, to meet the demands of consumers and investors and to compete effectively, decision makers must support intuition with supporting evidence.

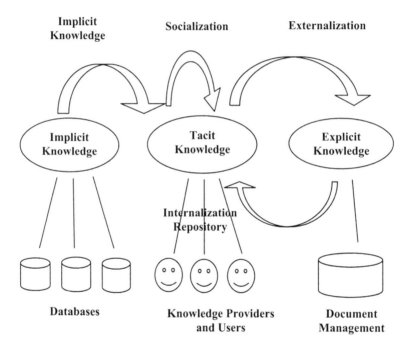

Figure 1. The Dynamic Nature of Knowledge Management

The information that provides such evidence is embedded in many different places: within corporate information systems, formulas in the on-line analytical processing (OLAP) systems, triggers in relational databases, visual patterns in geographic information (GIS) systems, predictive models in data mining packages, classification rules and personal profiles in text engine-based applications, business procedures in rules-automation systems, and drivers in decision-analysis tools (Thomsen, 1999). Successful knowledge management occurs when knowledge can be stored, accessed, verified, edited, and used through an enterprise-based data system that interfaces with the multiple data sources in traditional legacy databases.

Knowledge management deals with critical issues of organizational adaptation, survival, and competence in a rapidly changing environment. It embodies a synergistic combination of the data and information processing capacity of electronic technologies, and the creative and innovative capacity of human beings (Malhotra, 1998). Implicit in this description is a shift from an outmoded pragmatism to a more scientific approach. In the past, information served both control and prediction; it was used to meet the requirements of predefined organizational goals or best practices. To move ahead in these uncertain times, however, we must use information to stimulate questions rather than simply answers. What worked in the past may not work today or tomorrow. Thus, organizations must use innovative and creative approaches to collect and aggregate knowledge.

Knowledge can take various forms: experiential (tacit), literature-based, and evidence-based (both implicit and explicit). The bases of health care knowledge can be further defined as follows: Experiential or tacit knowledge is personal, is held in people's heads, and increases through "know-how" or by doing, following the modes of regular practice. This knowledge is informal, reflects daily management practices, and can be shared through socialization. Theoretical or literature-based knowledge is documented in reports and publications. Very often, it focuses upon a discipline, for example, the documentation of the structure-process-outcome relationship in health care. Theoretical knowledge offers frameworks to guide the development of evidence-based knowledge, both implicit and explicit.

In health care management, both implicit and explicit knowledge form the basis for evidence-based knowledge (Figure 1). (a) Implicit knowledge is derived by transforming a large quantity of electronic data into useful analytical charts or tables to support managerial decision-making. As such, implicit knowledge resides in databases, which may be constructed according to specific theoretical frameworks. Information applied to problem solving uses implicit knowledge. (b) Explicit knowledge is captured, documented, and stored in a data warehouse. It can be transmitted to share it among individuals. A complete problem solution can be imbedded in analytical files as a system to support decision-making. The media for this form of the knowledge are document-management system repositories that form such expert systems. When explicit, factual information is transformed into tacit knowledge, it is internalized. When tacit knowledge is applied to problem solving, it is externalized and made explicit.

In science, claims arise from theories, which are potentially fallible. The observations that lead to conclusions employ methods such as randomized trials or experimental procedures; as new information is gathered, beliefs that were once deemed certain may become obsolete. Scientific knowledge is the product of a collective modality of thought that is both refuted and confirmed by members of a scientific community (Malhotra, 1994). Consensus justifies causal connections and becomes the preeminent belief of the time. The problem is the potential for an entire body of scientific knowledge to be wrong. It was once believed that the earth was square. The goal of science is to ascertain the highest degree of consensus possible, with the hope that individuals are relying on proven facts and principles to derive conclusions. But consensus does not presume complete freedom from uncertainty. Nevertheless, the scientific method is the most objective method for collecting information.

Personal knowledge and experience often take precedence in health care managers' decision-making. Managers value past observations and anecdotal evidence from respected colleagues, which may be termed intuitive. Intuition is defined as the power to attain direct knowledge without interference and to exercise quick and ready insight. The process requires both awareness and utilization. Herbert Simon suggests that an effective manager does not have the luxury of choosing between intuitive and analytic approaches to problem solving (Brockman and Simmonds, 1997), but instead must balance both to solve each problem.

Many theorists have pointed out that observations are always interpreted in the context of prior knowledge. Kuhn states, "what a man sees depends both upon what he looks at and also upon what his previous visual-conceptual experience has taught him to see" (Malhotra, 1994). This tendency conditions health care managers to interpret different information in the same old fashion: the ideas that are familiar emerge and confirm current practices, while new ideas or less confirmed facts are pushed to the back burner. Health care executives are comfortable using intuition in their decision-making, but in the current volatile environment their focus must now include analytical evidence from scientific management studies, as well.

Evidence-based knowledge is derived from scientific replication and verification of facts. Used consistently and appropriately, it enables a manager to improve organizational performance. For example, mechanisms such as clinical case management, informatic integration, vertical integration, physician-hospital partnerships and patient- focused care may be used to integrate a health care delivery system. Although these mechanisms may have been shown to enhance a hospital's quality and efficiency, the way in which specific integration mechanisms operate independently or jointly to do so has yet to be demonstrated. Systematic investigations are needed to verify and replicate findings on the determinants of health care organizational performance.

One of the most promising avenues for organizational studies on performance uses data warehousing and multivariate mining techniques. Although data warehousing is now emerging as a valuable decision-making

4

tool, health care managers are realizing that the databases are only as useful as the quality, accuracy, and ease of use. Data warehouses organize information systems of interrelated factors such as organizational context, design, performance and outcome indicators. Analysts extract data from a myriad of locations, build a database that maintains and updates the data, and analyze the systems' data to provide meaningful information to users (www.dwinfocenter.org).

Data are valuable commodities. Managers constantly call for information to substantiate the workings of an organization. When data warehouses come into play, more information even is available than currently can be managed. They provide consistent, collaborative information that is easily accessible. The housed data consist of operational data, decision support data, and external/contextual data (www.techguide.com).

Many statistical approaches for mining the data are available (Berry, 1999; Kimball, 1998; Libowitz, 1999). The potential benefits of data mining in the health care industry are the myriad opportunities to establish benchmarks for continuous quality improvement and performance enhancement. Data mining can identify those interventions that are most successful, unsuccessful, or inefficient. It also can establish a benchmark for health care employees and can demonstrate their rankings. In terms of the quality of care and promoting a community's health, data mining can identify individuals or organizations that are at risk for adverse health events. Lastly, health care marketers can use data mining to develop and market new products and services. Such outcome management will weed out practices that do not serve the organization's best interest (Breshnahan, 1997; Cranford, 1998).

Such uses of information reflect a change in management philosophy. It was once assumed that only a few top-level health care executives who made decisions should be privy to information. As organizational structure changes and places middle managers in the role of decision makers, more individuals are seeking access to information. Today's organizations are large and complex, comprising many subsystems. The data warehouse's function is to extract data from these varying sites, using a group of programs to aggregate them into storage rooms or data marts (www.techguide.com).

As health care becomes a consumer-oriented business, there is a drive to "know" its customers: to forecast trends and identify questions that will help in managing customers. In the competitive, for-profit environment, knowledge of the customers is a strategy for survival. For organizational decision makers, data warehouses and analytic capacity promise increased speed and functionality. Using data warehouses connects internal organizational users of information with the multitude of outside users. That availability of information to the customers gives them a sense of connection and comfort with the organization's services. Using their information centers, health care organizations can identify customer values, evaluate products and services, develop contact strategy, and plan strategically (Stern, Bell, and McEachern, 1998).

Data warehouses can play a major role in evidence-based decision making (Kerkri et al., 2001). The question, however, is whether data warehouses are being designed with knowledge management in mind. The answer is no: most data warehouses are designed to create a "complete" product. If knowledge management were their driving principle, there would be no such end result; rather, data warehouses would be built on the premise of questions unasked and functions unknown. For the greatest usefulness, they should be built to meet the needs of an amorphous organization, one whose needs and interests may vary from quarter to quarter. The information supply chain must be organized to ensure that the right information is delivered to the right person at the right time. Furthermore, that information should initiate a cascade of actions, based on the information that adds value to the organization. The trick is to use information and knowledge as a product and integrate the process of acquiring it into the business process (www.dwinfocenter.org/casefor.htm).

By using evidence drawn from data warehouses, health care managerial decisions will promote growth and higher quality of care. The results should be better outcomes for patients and more effective performance for organizations. The problem at present, however, is that there is little literature on the correlation between evidence and the quality of decisions. Implementing evidence-based decision-making is difficult because of the highly fluid nature of management facing market forces in turbulent environments. The multitude of organizational and external factors makes "best practices" or standardization difficult to follow. Nevertheless, causal inference in health care management is a highly feasible way to achieve evidence-based knowledge, which can navigate an organization to high performance.

This book introduces the principles and methods for drawing sound causal inferences in research on health services management. The emphasis is on the application of structural equation modeling techniques and longitudinal analytic methods to develop causal models in health care management. Topics include causality, theoretical model building, and model verification. Multivariate methods and their applications in health care management are illustrated. They include path analysis, analysis of linear structural relationships (LISREL), meta analysis with latent variables, multilevel modeling, and growth modeling techniques.

The primary goals of the book are to present advanced principles of health services management research and to familiarize students with the multivariate analytic methods and procedures now in use in scientific research on health care management. The hope is to help health care managers become better equipped to use causal modeling techniques for problem solving and decision-making. More specifically, the objectives of the book are as follows:

1. To explain various causal-modeling techniques that may be useful in evaluating health care programs.
2. To develop the student's proficiency in applying measurement models to health services management research.
3. To clarify statistical problems in the application of structural equation and measurement models.
4. To review and critique the strengths and weaknesses of the above techniques as applied in health care management.
5. To identify health service management problems that are amenable to quantitative analyses.
6. To formulate theoretical models in health care management for empirical validation.
7. To employ knowledge management approaches to data warehousing and data mining.

The book has thirteen chapters. Following the introductory chapter, Chapter 2 discusses the foundations of causal inquiry in research on health services management. The third chapter explains how the structural relationships among organizational context, design, performance and outcome are established to guide data warehousing and data mining. The fourth chapter deals with the exploratory approaches to identifying structural relationships in research on health management. The fifth chapter introduces structural equation modeling and its applications to research on health care management. The sixth chapter illustrates the power of confirmatory factor analysis for validating the measurement models. The seventh chapter presents structural equation modeling with observed variables for both recursive and non-recursive models. The eighth chapter explains structural equation modeling with latent variables. The ninth chapter presents covariance structure models and explains how organizational theories can be validated and confirmed, using as an example a study of integrated care delivery systems. The tenth chapter shows how longitudinal panel data can be analyzed with multiple group analysis. The eleventh chapter introduces multilevel covariance modeling of organizational analysis of the determinants of job satisfaction and patient satisfaction. The twelfth chapter uses occupancy rates observed in the period of 1996-1999 as an example to assess the impact of the Balanced Budget Act on hospital operational efficiency, using growth modeling. Finally, the prospects for establishing integrated, evidence-based health care management in the quest for high performance are discussed.

REFERENCES

Berry, M.J.A. (1999). Mastering Data Mining: Art and Science of Customer Relationship Management. N.Y.: John Wiley and Sons.

Breshnahan, J. (June 15, 1997). A delicate operation. *CIO Magazine*.

Brockman E., Simmonds, P. (1997). Strategic decision making: the influence of CEO experience and use of tacit knowledge. *Journal of Managerial Issues* 9 (4): 454-468.

Cranford, S. (May, 1998). Knowledge through data mining. DM Review.

Ellrodt, G., Cook, D., Lee, J., Cho, M., Hunt, D., Weingarten, S. (1997). Evidence-based disease management. *Journal of American Medical Association* 278:1687-1692.

Gray, J.A.M. (1997). *Evidence-Based Health Care: How to Make Health Policy and Management Decisions*. London, Churchill Livingstone.

Kimball, R. (1998). The Data Warehouse Lifecycle Toolkit: Expert Methods for Designing, Developing and Deploying Data Warehouses. N.Y.: John Wiley and Sons.

Kerkri, E.M., Quantin, C., Allaert, F.A., Cottin, Y., Charve, Ph., Jouanot, F., Yétongnon. (2001). An approach for integrating heterogeneous information sources in a medical data warehouse. *Journal of Medical Systems* 25(3): 167-176.

Libowitz, J. (1999). *Building Organizational Intelligence: A Knowledge Management Primer*. N.Y.: CRC Press.

Malhotra, Y. (1994). On Science, Scientific Method and Evolution of Scientific Thought: A Philosophy of Science Perspective of Quasi-Experimentation [WWW document]. URL: http://www.brint.com/papers/science.htm.

Malhotra, Y. (1998). Knowledge Management, Knowledge Organizations & Knowledge Workers: A View from the Front Lines [WWW document]. URL: http://www.brint.com/interview/maeil.htm.

Sackett, D.L., Rosenberg, W.M., Gray, J.A., Hanes, R.B., Richardson, W.S. (1996). Evidence-based medicine: What it is and what it isn't. *British Medical Journal* 312:71-72.

Shortell, S.M., Zazzali, J.L., Burns, L.B., Alexander, J.A., Gillies, R.R., Budetti, P.P., Waters, T.M., Zuckerman, H.S. (2001). Implementing evidence-based medicine: the role of market pressures, compensation incentives, and culture in physician organizations. *Medical Care* 39(7): I62-I78.

Stern, L., Bell, L., McEachern, C. (1998). Data warehousing in the health care industry--three perspectives. *DM Review*. March.

Thomsen, E. (1999). *Microsoft OLAP Solutions*. New York: John Wiley and Sons.

www. dwinfocenter.org. A Definition of Data Warehousing [WWW document] URL:http:www.dwinfocenter.org/defined.htm.

www.techguide.com. Getting Started with Data Warehousing [WWW document] URL: http://www.techguide.com

www.techguide.com. Putting Metadata to Work in the Warehouse [WWW document] URL:http://www.techguide.com

www.dwinfocenter.org. The Case for Data Warehousing [WWW document] URL: http://www.dwinfocenter.org/casefor.htm.

CHAPTER 2

CAUSAL INFERENCE: FOUNDATIONS FOR HEALTH CARE MANAGERIAL DECISION-MAKING

INTRODUCTION

The scientific basis for managerial decisions is the use of evidence generated from explicit, experiential, and confirmed knowledge. This is a systematic thought process, beginning with the collection of observable facts and then moving to the analysis of these facts to arrive at an adequate explanation of the phenomenon under study. Ideally, scientific data should be gathered under a theoretically informed framework, so that evidence can be derived from the application of data warehousing and data mining techniques. Such evidence-based knowledge can be integrated with practical and experiential knowledge to shed light on the cause-and-effect relationships between the problems and solution sets in the field of health care management.

This chapter will explain the principles of causal analysis and will introduce multivariate analytic methods and procedures for conducting confirmatory studies. This chapter will also prepare readers to select appropriate study designs and causal-modeling techniques for problem solving and decision making in health services administration. Specifically, the chapter objectives are as follows:

1. To explain causal-modeling techniques that may be useful for evaluating health care programs.
2. To explain the measurement models of latent variables or constructs.
3. To examine the principles of causal analysis.
4. To review and critique the strengths and weaknesses of these techniques as they apply to the health care field.
5. To identify health services problems that are amenable to quantitative analyses.

This chapter has been adapted from my book chapter entitled "Assessing Causality Foundations for Population-Based Health Care Managerial Decision Making" in Denise M. Oleske (ed.), *Epidemiology and the Delivery of Health Care Services: Methods and Applications*, 2nd edition. New York: Plenum Press/Kluwer Academic Publishers, 2001.

Scientific Inquiry

This section will introduce the fundamentals of scientific inquiry, discuss a variety of research designs, introduce the principles and methods employed in causal analysis, and explain a variety of analytical designs and statistical approaches for identifying the determinants of health services problems. Scientific inquiry is used in all forms of problem solving, to support and verify or reject preconceived concepts and propositions about situations (Wan, 1995). At the lowest level, scientific inquiry helps to (1) describe a phenomenon and the facts that surround it. By identifying causal relationships, scientific inquiry also helps to (2) explain the phenomenon. (3) Predicting future events from the historical patterns of a phenomenon is another use of scientific inquiry. At the highest level, scientific inquiry can provide the information to (4) develop, modify, and design solutions to the problems presented (Figure 2).

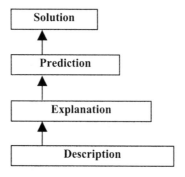

Figure 2. Foundation of Scientific Inquiry

Characteristics of Scientific Inquiry

Scientific inquiry has four characteristics: observability, verifiability, tractability, and manipulability. Each of these characteristics has a precise meaning for the process of scientific inquiry. Although all four elements are typically found in scientific investigation, they need not all be present in every case.

Observability is central to the empirical approach to the natural world: observations of sensory data that can be confirmed. Facts or data must be observable through either objective or subjective procedures. For example, health status may be identified through clinical assessment and laboratory results, which are objective measures of health. On the other hand, subjective measures such as surveys also provide observable measures of health status.

10

Verifiability is the second characteristic of scientific inquiry, and is closely related to observability. A phenomenon is considered verifiable when its existence or pattern is detected by repeated observations and is verified by diverse observers. An example is the historical trend of hospital data, which is used to estimate or forecast the use of health services. Care demand forecasting could be useless if not based on a large amount of sequential, time-series data. In another example, epidemiological studies stress replication, or repeated testing, in the search for pathogenic sources of health disorders.

Tractability of the data is a third characteristic of scientific inquiry. Useful data must be tractable or accessible. This term indicates the degree to which observed facts and data are both accessible and controllable. From an epidemiological perspective, a tractable disease is one that has traceable etiology, and is therefore preventable or controllable.

The final characteristic of scientific inquiry is *manipulability*, or the degree to which experiments can be performed on the observed subjects. For example, a genetic disease is considered manipulable if early detection and gene therapy are possible. The manipulability is precisely determined by explicit knowledge, so that predictable results can be achieved.

Logic of Scientific Inquiry

The ultimate goal of scientific inquiry is to generate a set of hypotheses that are testable through repeated observations and either confirmed or rejected. To derive, test, and confirm a theory requires the explication of concepts in a theoretical framework, as well as following the process of scientific inquiry. In health services evaluation and research, although in many instances existing theoretical frameworks are useful, other unique aspects of health care delivery and organizations lack a theoretical basis.

New theories are formulated by using two basic approaches: induction and deduction. The respective roles of these two logical methods of causal inference--for example, in epidemiology – have been hotly debated. Readers who are interested in the conflicting philosophical viewpoints about causal inference may review the 1985 proceedings of the annual conference of the Society for Epidemiologic Research (Rothman, 1988).

The inductive approach to theory construction starts by observing specific facts and then formulates general principles based on them. For example, the Joint Commission on Accreditation of Health care Organizations has set forth a series of activities for monitoring quality and improvement to enhance the performance of health care organizations. Quality indicators are collected routinely by assessing the presence of sentinel events (i.e., baby abduction, injuries due to treatment, and unanticipated deaths) or adverse patient outcomes (i.e., repeated hospitalization, treatment problems, and surgical complications). Then conclusions are drawn regarding which aspects of the health care process are associated with greater risks of adverse events. For example, specific observations, analyzed in terms of organizational theory, have suggested that poor clinical design of support systems and human errors may adversely affect the quality of patient care. Such generalization based on specific facts is called logical induction.

The deductive approach starts with general principles and derives specific expectations from them to be tested. Here, the theory or framework is the starting point from which specific hypotheses are proposed and tested empirically. In the previous example, investigators might choose a hypothesis before compiling data on sentinel health events. For example, if the theoretical framework chosen concerned the effects of care process design in a health care delivery system, one might hypothesize that case management with an automatic system for tracking patient outcomes would reduce mishaps, through care planning and coordination. Observations that case management reduces adverse events would support this hypothesis. Next, other data to correlate the case management strategies with patient care outcomes would have to be gathered, to either support or reject the hypothesis. In this way, a general principle or hypothesis about the effects of an automated outcome tracking system (i.e., a clinical decision support system) could be tested with specific observations.

Often the deductive and inductive approaches are used together; in fact, in scientific inquiry they may be inseparable. In actuality, observations of specific events may lead inductively to a conclusion and generalization about a phenomenon. Then, from this generalization, concepts and propositions may be formed and organized into an abstract theoretical structure. Using that theoretical structure, specific hypotheses can be proposed and tested by empirical data. The following is an illustration of the construction of a theory through both inductive and deductive processes.

Prevention of sentinel health events is a pressing quality-of-care issue for which existing organizational theory or epidemiological theory offers little guidance. The first step in constructing a theory about this problem is the observation of two specific phenomena: 1) Medication errors vary greatly by hospital units. 2) Those units with a clinical decision support system (i.e. automated prescription records) have lower annual medication errors than hospitals with no such system.

These interesting observations about the beneficial trend associated with an automated prescription record system might simply be happenstance. In order to postulate the relationship between these observations, they must be accurately defined and measured. Thus, the next step in the cycle of theory development is to measure the two variables, effective care process--in this case, prescribing medications--and medication errors.

After the two variables, prescribing behavior (care process) and medication errors, have been defined and measured, data about them must be collected. Then data analysis is conducted to identify any association or correlation. Analysis also measures the magnitude of any relationship between the two variables. Furthermore, the effect that a change in one variable may have on the other can be examined using predictive analytical techniques. In this step, any relationship between the two variables is defined and described.

This example began with a focus on specific data about an observed phenomenon. After data analysis, generalizations, or tentative explanations about the relationship between the two variables, might include:

1. Effective use of an automated prescription record system leads to a lower rate of medication errors.
2. Use of an integrated clinical care process varies directly with the size of hospitals.
3. Integration of clinical care is more likely if an automated system for patient care has positively affected the efficiency and effectiveness of patient care.
4. Hospital performance varies directly with the degree of clinical and informational integration.

These generalizations describe relationships between the use of care technology and hospital performance, as defined within a health system framework (Wan, 1995).

From such generalizations, founded in observation and empirical data analysis, it is possible to move into a deductive mode, in which *propositions* can be formed. Propositions, or statements about the relationships between concepts, are an important part of the theoretical framework, because they can be tested as hypotheses.

Two propositions that could be deductively derived here are:

More clinical and informational integration leads to higher efficiency in care processing, and in turn reduces annual rates of adverse patient outcomes.

Clinical and informational integration has enhanced hospital performance because the design and use of care technology offer better decision support and reduce undesirable outcomes.

These theoretical propositions, describing possible relationships between the two concepts, constitute hypotheses that can be tested. They return us to the point of making observations about relationships in the real world, and complete the circle of observation-deduction-concept formulation-hypothesis testing-observation.

Stages in Scientific Inquiry

When designing solutions for a problem in health care management, the initial and most crucial decision is to define the condition or problem to be examined. Epidemiological surveys for needs assessment are particularly useful in defining the problem, after which an intervention can be designed. Then evaluation research methods appropriate to that intervention design and measures for quantifying the criteria can be selected. This detailed process, which follows the stages of scientific causal inquiry, consists of seven steps (see Figure 3): 1) identification and specification of the study problem, 2) selection of a theoretically informed framework to guide the research process, 3) quantification of the study variables, 4) specification in analytical modeling, 5) selection of an intervention design and analysis, (6) confirmatory analysis, and (7) establishment of causality through replications. These steps will be described in this section.

1)   Identification & Specification of the Study Problem

2)   Selection of an Informed Theoretical Framework

3)   Quantification of the Study Variables

4)   Specification in Analytical Modeling

5)   Selection of the Intervention Study Design & Analysis

6)   Confirmatory Analysis

7)   Establishment of Causality

Figure 3. The Seven Steps of Scientific Causal Inquiry

Once overall research questions have been developed, it is necessary to state the study problem in clear, precise terms and to specify its salient aspects. A useful problem statement will guide investigators in the design and execution of quantitative root-cause analysis.

1. Identification and Specification of the Study Problem

The starting point for this process is to identify key issues and variables in the problem. Every problem affects three dimensions: persons, time, and place. In other words, the persons involved, the place or location, and the time element or causal chains must be clearly identified. Then the study problem must be specified, which involves elucidating its precise attributes. Attributes of particular importance are the problem's magnitude and its significance, its location and boundaries, and its determinants and consequences. Magnitude and significance refer to two important statistics: the incidence of the problem and the prevalence of the problem. The incidence of a problem is the number of new episodes of the problem that occur within a certain time interval. Prevalence is the number of old and new episodes of a problem existing at a given point in time. Incidence refers to the timing of the problem; prevalence refers to the sheer number of the events. Considered together, these two aspects of the problem can estimate the magnitude of its seriousness.

To understand and interpret these attributes clearly, the specification process must provide operational definitions. An operational definition describes a variable or concept in terms of the procedures by which it can be measured. For example, if the study problem is an examination of whether patients are being discharged too soon, the variable of early discharge could be defined as "discharged on the first or second post-operative day."

2. Selection of an Informed Theoretical Framework

Theories can be defined as abstract generalizations that present systematic explanations about the relationships among phenomena; theories also knit together observations and facts into an orderly system (Bagozzi, 1980; Polit and Hungler, 1987; Mulaik, 1987). A theoretical framework is a statement by the researcher of the assumptions and beliefs that guide a particular research process. Theory also provides an analytical framework that can help logically interpret the facts collected in a study, and it guides the search for new information. Selecting a theoretically informed framework consists of five stages: (1) conceptualization, (2) model selection, (3) critique of previous work in the field, (4) review of evidence, and (5) refinement or reformation of the model.

The first stage, conceptualization, identifies key elements of the problem in relation to clearly established principles. The investigator formulates the study problem in abstract, conceptual terms. This formulation will ultimately help to explain the data. It is important to remember, however, that conceptual thought is a deductive process, and so requires testing in the real world.

After the relevant concepts are separately distinguished, the next stage (2) is to develop a theoretical model of the study problem. The model gives the real-world attributes and manifestations of the study problem an abstract representation. Portrayed that way, relationships between variables can be examined readily and analyzed. This abstract representation of variables and relationships is the basis of hypotheses, which are proposed later in the research process.

Once the investigator has reformulated the study problem in abstract terms, it is helpful to examine work by other researchers or managers on the same or similar issues. Such a review of the literature (stage 3) can include research findings, theory, methodological information, opinions, and viewpoints, as well as anecdotes and experiential descriptions. Often there is a wealth of available literature, and the investigator must decide how relevant particular material is to the study problem.

After the review of the pertinent literature, the evidence that led to the selection of the study problem must be reexamined (stage 4). For example, if the quarterly audit of patient incident reports at a general hospital indicates that there has been an increase in patient falls, it would be useful to compare data from that audit with information in the literature. For example, while the literature suggests that elderly patients are high-risk candidates for falls while in the hospital, the audit data might show that the majority of patients who fell were middle-aged, post-operative patients who were receiving narcotics for pain relief. Thus, both age and the effects of medications should be included as variables in a theoretical framework for a management study about the causes of patient falls. In addition, any confounding factors that might distort or suppress the relationship between an intervention variable and a response variable should be considered in the formulation of a model. Such an analysis ties the development and choice of a theoretical framework in the management study to existing experimental data and the real world.

3. Quantification of Study Variables

A variable can be defined as a characteristic of a person or object that varies within the population under study. The problem statement and the theoretical framework for a confirmatory study will have identified variables. Next, if the results of the study are to be meaningful to the investigator and communicated clearly to users, they must be defined clearly. This clarity of communication is achieved by using operational definitions and measurement techniques. The study variables in health care research can be measured at the individual (patient) or aggregate (organization or community) level (Marks, 1982).

An operational definition links an abstract concept found in the problem statement and the theoretical framework to a variable that can be measured and quantified. For very broad concepts, proxy measures are used; for example, patient satisfaction is often used as a proxy measure for quality, which is a more ambiguous concept. Once the variables have been operationalized, propositions stating the relationships between the variables are presented as theoretical hypotheses.

Important steps in operationalizing a variable include specifying how the variable will be observed and measured. Variables differ considerably in the ease with which they can be described, observed, and measured. Even something as simple as body weight can be measured in pounds or kilograms, as well as in fractions. In addition, body weight may show diurnal variation, which suggests that the time of day when the weight is measured should be specified in the operational definition. Furthermore, body weight may also be examined using anthropomorphic measurements to determine the percentages of lean muscle mass and body fat.

To explore the example further: an operational definition of body weight depends on the problem under study and where it fits in a theoretical framework. If the problem concerns overall population health, a weight to the nearest pound is adequate. In a weight reduction clinic, on the other hand, detecting small weight losses might be of moments, and so the operational definition should specify weights taken first thing in the morning. Another example would be research by an exercise physiologist studying long-distance runners, which would require more precise anthropomorphic measurements.

This example of body weight research has described a number of options for linking that concept with measurement. The most precise observations and measurements must be chosen to describe the attributes, magnitude, and significance of variables, as well as to identify causal relationships between them.

Beyond that, however, the validity and reliability of the measurement techniques used to arrive at an operational definition must be established. Only if a measurement is both valid and reliable can it be depended upon as an accurate representation of a concept or attribute. Validity is the degree to which the tool or technique actually measures what it is supposed to measure. Reliability is the extent to which a tool or technique will measure the same response every time it is used to measure the same thing. Furthermore, validity and reliability cannot be considered as independent qualities; a measuring device that is not reliable cannot possibly be valid. Validity and reliability are important criteria both for existing measurement tools and for the development of new instruments.

4. Specification in Analytical or Causal Modeling

The causal analysis, which will be guided by the study questions and the theoretical framework, uses statistical modeling techniques to simplify, summarize, organize, interpret, and communicate numerical data. One builds a model, which is assumed to describe, explain, or account for the data in terms of relatively few parameters. The model is based on a priori information about the data structure (whether from theory or hypothesis, design, and/or knowledge from previous research). On the basis of the available data, one wants to test the validity of the model and to test hypotheses about the parameters of the model. In evaluating treatment outcomes, researchers often need to analyze multiple outcomes (e.g., the complication rate, repeated hospitalization rate, and hospital mortality rate). Sometimes these outcome variables are correlated with each other. In that case, it is imperative to use a multivariate statistical technique to examine the effects of an intervention on multiple outcome variables, with and without correlated errors or residuals (Short and Hennessy, 1994). The outcome variables are treated as endogenous variables, and the intervention variable is treated as an exogenous variable.

The specification of causal links among multiple study variables is not a simple matter. Bollen (1989) states that no single definition of causality has emerged that is routinely employed in sciences and other fields. However, the explication of causal links in terms of structural relationships is useful in the search for causality. With an informed theoretical framework, investigators can map out ideas, theory, and hypothesized relationships, and can portray structural or causal relationships among the study variables. Bollen identifies three components of cause: isolation, association, and the direction of influence (1989). Criteria for specifying cause and effect relationships are direction, temporal sequence, strength of the association, substantive meaning of the structural relationships, and verifiability of the cause-effect relationships.

In research on preventive care, Kraemer et al. (1999) suggest using a potency measure of risk or predictor factors. Once a risk factor is assessed for an effect size that is interpretable and meaningful in a clinical or policy context, it can be selected as a pertinent predictor variable for the variation in a dichotomized outcome variable (e.g., risk of being hospitalized). Sensitivity and specificity tests, using a specific risk factor, can be computed for this outcome variable. The potencies of the risk factors can be relatively determined by observing the height of "equipotency curves". The higher the observed curve, the more potent the risk factor for predicting an outcome. This approach can help establish refined criteria for selecting risk factors for causal inference.

5. Selection of an Intervention Study Design & Analysis

An important point to understand is that the evaluation of a program depends on how the program intervention is designed. The first and most crucial decision when designing an intervention is to define the condition or problem it will address; in defining the magnitude of the problem, in particular, epidemiologic surveys for needs assessment are useful. Next, evaluation methods appropriate to that intervention design can be selected, and the measures for quantifying the evaluation criteria specified.

In assessing a program, researchers make three basic assumptions: 1) the intervention has been effectively designed and the integrity of the treatment has been maintained; 2) that the outcome (performance) measures, which may be single or multiple, are predetermined by the goals set when the program was planned; and 3) the measurement of specific outcome(s), using either an aggregate, single index or multiple indicators, are logically formulated and can be empirically validated by the analysis; these outcome indicators should be repeatedly measured. In evaluating programs, valid and reliable indicators measured at multiple time points are essential, because this is the basis for accurate causal inference.

The above assumptions guide the selection of measures. The structural, process and outcome criteria previously described by Donabedian (1982), also apply. An intervention can be measured by a discrete variable (presence or absence of an intervention) or by a continuous variable (the amount/magnitude of intervention). When evaluating populations, these measures are usually rates, or odds ratios (the likelihood of events in relation to exposure to an intervention).

A researcher uses either an experiment with a single intervention variable, or a complex factorial design with multiple interventions. Similarly, program performance is measured either by a single (or aggregate) outcome indicator or by a set of correlated outcome indicators. Following the specification of intervention and outcome measures, appropriate statistical methods are used to analyze program performance (Table 1). Detailed descriptions of the types of study designs and statistical outcome analysis follow the table.

The next subsections describe the types of study designs and the statistical methods for assessing outcomes.

Cross-Sectional Study. A cross-sectional study is a simple form of a correlational design. In this design, all measurements are taken at one point in time. Due to its simplicity and ease of administration, the design is popular in health service research when researchers have little control over observations. The cross-sectional design is useful for deciding whether two or more variables are related. Usually identification of relationships is the extent of the study question. It is important to remember that correlation does not imply causality.

Table 1. Statistical Methods for Analyzing Program Performance, by Study Design and Type of Outcome

Study Design	Analysis of Program Performance	
	Single Outcome	Multiple Outcomes
Single Intervention (Classical experiment or quasi-experimental design)	OLS regression analysis Logistic regression model Proportional hazards model	Linear structural relations (LISREL) model
Multiple Interventions (Factorial design or quasi-experimental design)	Analysis of variance Single or multiple time-series analysis Proportional hazards model	LISREL model Meta-analysis with latent variables

Time-Span or Longitudinal Study.   (a) A *retrospective* study examines past events and their effects on the present situation (Breslow and Day, 1980; Szklo and Nieto, 2000). For example, hospitals with service delivery problems, which a study would term the cases, are compared to those without the problem, the controls. Past data on the two groups are compared for the use of a specific support system for managerial decisions, such as an integrated clinical and information management system. Differences in the number of incidents of the study problem are examined to see whether they can be attributed to the presence or absence of such a mechanism.

(b) *Prospective study*. This design is also called a population-based study design (Green and Wintfeld, 1993; Lakka et. al., 1994; Lilienfeld, 1976; Murray, 1998; Norell, 1992). The normal subjects, either individuals or organizations, are selected; then some are exposed to a risk or a factor of organizational system design. The outcomes are then assessed over time. If significant differences in outcome measures are observed between those exposed and those not exposed to the study treatment, the researcher can conclude that a cause-effect relationship exists between that risk or organizational design factor and organizational performance.

(c) *Classic Experimental Design*. This design uses randomization, which assigns people, organizations, departments, etc., to treatment or control categories. This study design is commonly called a pretest-posttest control group design. The design is shown diagrammatically below:

$$\frac{R\,O_1\,X\,O_2}{R\,O_1\quad O_2},$$

where:
R refers to the random assignment;
O_1 refers to the pretest results;

O_2 refers to the posttest results;

X is the intervention program (treatment).

Under perfect conditions, this design enables the researcher to generate precise information about the contribution of an intervention variable (X) to the difference in the measure of an outcome or response.

(d) *Complex Experimental Evaluation Design.* Since evaluation looks for effects that can be attributed directly to program interventions, evaluations frequently include control variables, to determine net changes in the outcome variable by simultaneously controlling the effect due to exogenous factors. Unfortunately, however, a perfect statistical control of the influence of exogenous variables is not feasible. An alternative method is to gather baseline data so that they can serve as their own controls. That approach, however, does not satisfy the need to measure and separate exogenous influences. One is left with the problem of balancing control and baseline information appropriately. Use of control areas multiplies the requirements for data, yet may be essential; accordingly, the control-baseline balance must be judged case by case.

A complex experimental design can handle many of the confounding effects of exogenous factors on the program outcome(s) or performance. In addition, it can analyze the joint, or interaction, effects of multiple causal variables on the outcome. For example, a factorial design can be used to examine the effects of several different types of intervention variables on the outcome variable(s). Factorial designs examine these variables simultaneously in the same experiment. This approach is economical and provides sound data. The statistical model for a two-factor experiment is as follows: $Y = f(x_1, x_2, x_{12}) + \text{errors}$. The total effect is the sum of mean effect, main effects (linear effects), an interaction effect (x_{12}), and errors. Analysis of variance technique can be employed to examine the additive and interaction effects of the two interventions on a response variable (Wan, 1995).

(e) *Quasi-experimental Design.* This design applies an experimental model of analysis and interpretation to bodies of data not meeting the full requirements of experimental control. Although many planned interventions in the field of health care have been carried out with classic experimental designs, others have lacked procedures in which it was possible to have complete experimental control or randomization of treatment. Quasi-experimental designs include, for example: time-series design, matched groups design, and regression-discontinuity design. (i) A time-series design is a sequence of numerical data in which each datum is associated with a particular instant in time (Ostrom, 1978). A series of measurements provides multiple pre-tests and post-tests. For example: O_1 O_2 O_3 O_4 x O_5 O_6 O_7 O_8. If, in this series, $O_5 - O_4$ shows a significant change

21

(difference), then maturation, testing, and regression effects are shown not to be plausible, since they would predict equal or greater changes (e.g., $0_2 - 0_1$).

(ii) The matched-group series design operates on the same principles as those described for the single-time series, except that a comparison group is matched with the treatment group to the extent possible, as follows:

Experimental Group: 0_1 0_2 0_3 x 0_4 0_5 0_6

Comparison Group: 0_7 0_8 0_9 0_{10} 0_{11} 0_{12}

In this design, multiple observations are collected on both experimental and comparison groups, before and after the introduction of the intervention (x). This design is most useful when no cyclical or seasonal shifts are observed in the investigation. The statistical analysis can be performed for the data by employing a paired-t test. The design does not randomly select the subjects for the experiment. Rather, it uses a set of criteria selected by the evaluator to give an intervention (for example, awards) to one group (experimental) and not to another group (comparison group). The performance of each group is then assessed by the evaluator. Two regression analyses are performed to determine the intercepts. The difference in the intercepts between the two groups can be used as a measure of, for example, the intervention effect on productivity. This approach can serve as an alternative method for outcomes research because it can improve the assessment of the effect of a program intervention (Trochim, 1990; Luft et al., 1990).

The statistical methods for analyzing program performance, or outcomes, are presented below.

Multiple Logistic Regression Model. The multiple logistic regression model is used when the dependent variable is measured by a binary or discrete variable, and the independent variables (risk factors or interventions) are continuous and discrete variables. Since the dependent variable is a discrete variable (e.g., the probability of being hospitalized in a specified period), the predicted probability should lie in the unity boundary. Logistic regression is preferable to ordinary least squares (OLS), because OLS estimates are biased and yield predicted values that are not between 0 and 1.

Structural Equation Modeling. In evaluating the performance of health care organizations, researchers often need to analyze multiple outcomes (e.g., the surgical complication rate, medication error rate, treatment problem rate, etc.). Sometimes these negative outcome variables are correlated with each other (Wan, 1992; Wan, Pai, and Wan, 1998). In that case, a multivariate statistical technique should be used to examine the effect of an intervention on multiple outcome variables, with or without correlated errors or residuals. The outcome variables are treated as endogenous variables, and the intervention variable is treated as an exogenous variable. In addition, confounding or control variables can be included in the model specification so that the net effect of an exogenous variable on the endogenous variable(s) can be ascertained (Al-Haider and Wan, 1991; Biddle and Marlin, 1987).

The analysis of linear structural relationships (LISREL) among quantitative outcome variables is useful in data analysis and theory construction. The LISREL model has two parts (Bollen, 1989; Hayduk, 1987; Jöreskog and Sörbom, 1979; Long, 1983; Maruyama, 1998). One is the *measurement* model, which specifies how the latent variables (e.g., adverse patient outcomes) or hypothetical constructs are measured by observable indicators (e.g., medication error rate, complication rate, patient fall rate, etc.). The other is the *structural equation* model, which represents the causal relationships among the exogenous and endogenous variables. The statistical interaction terms of exogenous latent variables can be incorporated in the model for assessment (Jaccard and Wan, 1996). Furthermore, longitudinal data can be used to validate the measurement and structural equation models (Boles and Wan, 1992; Hays, et al., 1994; Reed, 1998).

6. Confirmatory Analysis

Inferential statistics are used to draw conclusions about the study group, and to make generalizations about a large class of subjects on the basis of data from a limited number of subjects. Although we are accustomed to generalize or draw conclusions about a larger group through ordinary thought processes, inferential statistical methods provide an objective, systematic framework, so that researchers working with identical data will be likely to draw the same conclusions. Examples of inferential statistical methods are estimation of population parameters, hypothesis testing, tests of statistical significance, and identification and estimation of sources of variance or error.

An important part of data analysis is comparing the study results to the proposed hypotheses. Hypothesis testing is how researchers make objective decisions about the results of a study. Using statistical methods, the researchers can decide whether the study results reflect chance differences between groups, or true differences.

Hypothesis testing follows the rules of negative inference, in which a contradictory, or null, hypothesis is proposed. The null hypothesis states that there is no relationship between the variables, and that any observed relationship is due to chance (Biddle and Marlin, 1987; Wan, 1995). For example, assume that an investigator believes that a certain drug is effective in treating a specific disease. The process of determining if the drug is indeed helpful would begin with the null hypothesis (H_0) that the drug does not treat the disease, but that the illness is cured for reasons other than the use of the drug. Tests would then show whether the null hypothesis is correct (accept H_0), or whether it is incorrect and the drug has an effect in treating the disease (reject H_0). This rejection or acceptance of the null hypothesis is determined through statistical analysis of the study data.

The two types of errors in testing a hypothesis (Table 2) are: rejecting a null hypothesis as false when it is actually true (Type I error) and accepting a null hypothesis as true when it is actually false (Type II error). The term, level of significance, is used to describe the probability of committing a Type I error, and this can be controlled by the researcher; the level of significance is also referred to as alpha (α). The probability of committing a Type II error is beta (β), and its complement, $(1-\beta)$ is referred to as the power of a statistical test.

Table 2. Type I and Type II Errors in Hypothesis Testing

	H_0 is true	H_0 is false
Accept H_0	No error	Type II error or β error
Reject H_0	Type I error or α-error	No error

Hypothesis testing verifies the model that is specified in the theoretical framework, through procedures that analyze the attributes and relationships of the variables described in the study problem. For example, if a study on the utilization of emergency services examines whether certain age groups are more likely to use certain types of services, one hypothesis might be that young adults, middle-aged adults, and the elderly have similar requirements for emergency services. Suppose that descriptive statistical analysis of the data on young adults, middle-aged adults, and the elderly shows that young adults and the elderly have similar utilization patterns for treatment of skeletal trauma and head injuries, but that the main cause for young adults is motor vehicle accidents, and that for the elderly it is falls; while middle-aged adults' emergency service use is more for medical problems such as myocardial infarction and hypertension. Using inferential statistics, the investigator would identify what were the sources of variation and determine whether the differences were statistically significant. Furthermore, the goodness of fit statistics of a hypothesized model would indicate the usefulness of a theoretical model. This confirmatory approach does not necessarily reveal the truthfulness of the causal model in reality, but only the extent to which a hypothesized model can be nullified. In many cases, an alternative model with more carefully specified causal paths or study variables can help reduce the residual error and can search for additional sources of error (Bullock, Harlow and Mulaik, 1994; Szeinbach, Barnes, and Summers, 1995).

7.   Establishment of Causality

The notion of causality applies whenever the occurrence of one event is reason enough to expect the production of another.   Principles of causality are:

Causal Ordering: The presumption that one event causes another requires that the first event (X_1, X_2, or X_3) produce an expectation for the occurrence of the second event (X_4). (See Figure 4.)

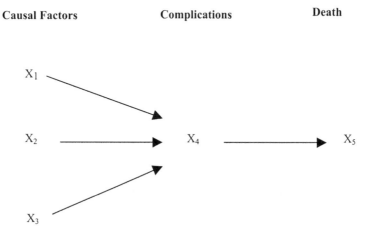

Figure 4. Causal Ordering with Regard to Disease

Temporal Ordering: A causal factor (X1 or X2) precedes its effect (X4). The causal direction assumes that an asymmetric relationship between the two study variables exists. (See Figure 4.)

Strength of the Association: The association does not necessarily imply causality. However, the stronger the association between the two variables, the more it supports a causal link between them.

Specification of Structured Process: A causal relation is assumed and validated by examining whether or not one event directly causes another if an effective operator is available to support the relationship. The specificity of an association entails a description of the precision with which the occurrence of one causal variable, to the exclusion of other factors, will predict the occurrence of an effect (Weed, 1988). The linear model assumes a unidirectional cause-effect relationship. However, in health care, problems do not always match that conventional pattern: variables are often related in reciprocal causation (see Figure 5).   For example, use of inpatient care services (Y1) and use of outpatient care services (Y2) are two endogenous

25

variables that are influenced by a selected group of predictor variables such as perceived health status (X1), age (X2), and social class (X3). Reciprocal causal links or paths can be established between the two endogenous variables such that the increase in inpatient care is affected by the increase in outpatient care, or vice versa.

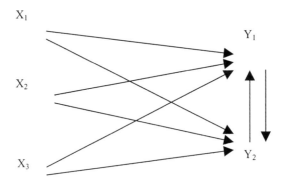

Figure 5. Mutual Causation, or Reciprocal Causal Paths

Verification of the Cause-Effect Relationship: The logical consistency of the tested model is a rigorous criterion for determining the plausibility of causality in a study. Obtaining repeated results under the same constraints can also verify the usefulness of the proposed analytical or causal model. The adequacy of statistical control for the extraneous influences is a critical requirement for causal inference. Control involves holding the variables constant or varying them systematically so their effects can be phased out from a study or compared to other conditions. Control may involve the active manipulation of subjects or conditions by researchers, or simply the structure of a study and the manipulation of data (Spector, 1981).

In experimental study designs, control usually refers to holding constant the level of a variable. Thus, a researcher may hold constant a particular variable, such as teaching status of hospitals, or sex and age of patients. In non-experimental study designs, control may be achieved by deliberately deleting observations that do not have the characteristics of the study. A researcher interested in urban hospitals may remove all rural hospitals from the study population.

There are limits to the degree of control that a researcher can exert on a study. For example, if there is a relationship between an independent variable and the dependent variable such that one cannot be held constant without the other, the variables are said to be confounded. Control is holding constant those variables that are not of direct interest in the current research, so they do not contaminate or confound results. However, studies limited to certain values of a control variable may not be generalized to other values--e.g., studies conducted on hospitals may not be generalized to nursing homes.

Substantive Meaningfulness of the Cause-Effect Relationship: Because the causal links are theoretically based and empirically validated, greater certainty in causal inference relies on coherence of the causal interpretation (Bullock, Harlow, and Mulaik, 1994; Susser, 1973) that corroborates with existing knowledge, and confirmation by the model fit statistics. However, since statistical results may be biased for various reasons, such as a biased sample, small sample size, and homogeneity of the study population, the meaningfulness of the postulated causal relationships must be supported by a coherent logic and interpretation.

Predictability of the Causal Relationship: If the occurrence of a specific outcome or effect can be predicted with great certainty when a direct causal factor is present, the causal model can be said to have predictive power. Predictability is an important ingredient of causal inference (Rothman, 1998).

Simulation of the Results: Modeling strategies vary by the types of data and designs. One can use data mining techniques to simulate the model to obtain optimal solutions. However, in actuality there is no best way to fit the model, as Bollen (2000) points out. In comparing alternative models with constraints, one may generate results that shed some light on the plausible causal relationships between exogenous and endogenous variables.

Meta Analysis with Latent Variables: Analysts often encounter the difficulty in drawing conclusions from multiple studies, because the results generated from independent studies are inconsistent. The art or science of combining evidence from different studies on the same study problem is based on the application of statistical procedures to collect empirical findings from independent studies for the purpose of integrating, synthesizing, and making sense of the results. This approach is referred to as meta analysis or epidemiology of results. Meta analysis can be used to assess risk factors (Allison, et al., 2000; He et al., 1999) and program outcomes (Rosenheck, 2000). For program evaluation, meta analysis may be used to estimate the effect size (ES), or intervention effect, of specific program outcomes. If it is properly used, meta analysis can strengthen the causal interpretations of non-experimental data (Cordray, 1990; Takkouche, Cadarso-Sqarez, and Spiegelman, 1999; Greenland, 1994). Evaluation researchers can develop a covariance structural model for multiple, comparable studies and then empirically test the model's goodness of fit by using meta analysis.

The current approach to knowledge management stresses the use of data warehousing and data mining strategies to organize a meta database for simulating analytical results (Liebowitz, 1999). There is a distinct possibility that researchers can apply meta analytical techniques to confirm or disconfirm a theoretical model that implies a causal linkage between exogenous and endogenous latent variables.

27

Multilevel Analysis

A data set that comprises patient, nursing unit, hospital, and community attributes or variables constitutes multilevel data. When multi-level data are used in a study, researchers tend either to disintegrate the aggregated data to the lower level, i.e., assign the value of aggregated data to the lower level; or to aggregate the lower level data to the upper level, i.e., use the mean or median of the lower level to represent the measurement of the upper level. From the methodological viewpoint, the former cannot satisfy the assumption of the independence of observations that underlies the traditional statistical approach (Bryk & Raudenbush, 1992; Duncan et al., 1998; Heck and Thomas, 2000). Another problem posed by disaggregation is that statistical tests involving the variable at the upper-level unit are based on the total number of lower-level units, which can influence estimates of the standard errors and the associated statistical inference (Bryk & Raudenbush, 1992; Hoffmann, 1997). The aggregation may lose valuable information, in that the meaningful lower level variance in the outcome measure is ignored by the process of aggregation (Hoffmann, 1997). Aggregation also may cause the "ecological fallacy", i.e. analyzing upper level data but interpreting the result at the lower level. In fact, most data are hierarchical; for example, patients are nested in nursing units, nursing units are part of a hospital, and a hospital is nested in a county or city. The hierarchical nature of data should not be neglected in either theory building or data analysis (Kaplan and Elliott, 1997; Muthén, 1991; Muthén, 1994; Phillips et al. 1998).

Multilevel analysis is frequently used when contextual and/or ecological variables are involved in the investigation of individual patient differences. It can be used to identify contextual effects and to derive accurate estimates of individual (patient)-level effects on a health care outcome variable (Morgenstern, 1998; Little, 2000). The multilevel model is an important step to tease out the net effect of each individual-level predictor variable on an outcome variable when the effects of ecological predictors, other individual-level predictors, and their interaction terms are simultaneously controlled.

CONCLUSION

This chapter has identified and explained the fundamental principles for conducting causal analysis in health services management. Epidemiological study designs and methods were reviewed to show how multi-causal factors associated with managerial problems can be analyzed. The application of knowledge management to health services problems can guide the development and implementation of solution sets--namely, interventions. The principal criteria of causality and their application in causal analysis are essential to developing scientific, evidence-based knowledge to guide organizational changes (Keats and Hitt, 1988) and innovations (Scott and Bruce, 1994).

The causal analysis advocated here does not imply that employing explicit and practical knowledge can solve every managerial problem. However, by applying causal analysis, managers may be able to search more efficiently for errors that are amenable to organizational and behavioral interventions.

REFERENCES

Al-Haider, A.S., Wan, T.T.H. (1991). Modeling organizational determinants of hospital mortality. *Health Services Research* 26(3): 303-323.

Allison, J.J., Kiefe, C.I., Weissman, N.W., Person, S.D., Rousculp, M., Canto, J.G., Bai, S., Williams, O.D., Farmer, O.D., Centor, R.M. (2000). Relationship of hospital teaching status with quality of care and mortality for Medicare patients with acute MI. *Journal of American Medical Association* 284: 1256-1262.

Bagozzi, R. P. (1980). *Causal Models in Marketing.* New York, NY: John Wiley and Sons.

Biddle, B.J., Marlin, M.M. (1987). Causality, confirmation, credulity, and structural equation modeling. *Child Development* 58: 4-17.

Boles, M., Wan, T.T.H. (1992). Longitudinal analysis of patient satisfaction among medicare beneficiaries in different HMOs and Fee-for-Service Care. *Health Services Management Research* 5: 198-206.

Bollen, K.A. (1989). *Structural Equations with Latent Variables.* New York: John Wiley & Sons.

Bollen, K.A. (2000). Modeling strategies: In search of the holy grail. *Structural Equation Modeling* 7(1): 74-81.

Breslow, N.E., Day, N.E. (1980). *Statistical Methods in Cancer Research. Volume I: The Analysis of Case-Control Studies.* Lyon, Switzerland: International Agency for Research in Cancer.

Bryk, A.S., Raudenbush, S.W. (1992). *Hierarchical Linear Models.* Newbury Park: Sage Publications.

Bullock, H.E., Harlow, L.L., Mulaik, S.A. (1994). Causation issues in SEM research. *Structural Equation Modeling* 1(3): 253-267.

Champagne, Francois (October 1-3, 1999). The Use of Scientific Evidence and Knowledge by Managers. Paper presented at the 3rd Conference on the Scientific Basis of Health Care, Toronto, Canada.

Corday, D.S. (1990). Strengthening causal interpretations of nonexperimental data: The role of meta analysis. In L. Sechrest, E. Perrin, and J. Bunker (eds.) Research Methodology: Strengthening causal interpretations of nonexperimental data. Rockville, MD: Agency for Health Care Policy and Research.

Donabedian, A. (1982). *Explorations in Quality Assessment and Monitoring. Volume II. The Criteria and Standards of Quality.* Ann Arbor, Michigan: Health Administration Press.

Duncan, T. E., Alpert, A., Duncan, S.C. (1998). Multilevel covariance structure analysis of sibling antisocial behavior. *Structure Equation Modeling* 5(3): 211-228.

Green, J., Wintfeld, N. (1993). How accurate are hospital discharge data for evaluating effectiveness of care? *Medical Care* 31: 719-731.

Greenland, S. (1994). Invited commentary: A critical look at some population meta-analytic methods. *American Journal of Epidemiology* 140 (3): 290-296.

Hayduk, L. A. (1987). *Structural Equation Modeling with LISREL.* Baltimore: Johns Hopkins University Press.

Hays, R.D., Marhall, G. N., Yu, E., Wang, I., Sherbourne, C. D. (1994). Four year cross-lagged associations between physical and mental health in the medical outcomes study. *Journal of Consulting and Clinical Psychology* 62(3): 441-449.

He, J., Vupputuri, S., Allen, K., Prerost, M.R., Hughes, J., Whelton, P.K. (1999). Passive smoking and the risk of coronary heart disease—A meta-analysis of epidemiologic studies. *New England Journal of Medicine* 340(12): 920-926.

Heck, T., Thomas, S. (2000). *An Introduction to Multilevel Modeling.* Mahwah, New Jersey: Lawrence Erlbaum Associates, Publishers.

Hoffman, D.A. (1997). An overview of the logic and rationale of hierarchical linear models. *Journal of Management* 23(6): 723-744.

Jaccard, J., Wan, Choi K. (1996). *LISREL Approaches to Interaction Effects in Multiple Regression.* Thousand Oaks, CA: Sage Publications.

Jöreskog, K.G., Sörbom, D. (1979). *Advances in Factor Analysis and Structural Equation Models.* Cambridge, MA: ABT Books.

Kaplan, D., Elliott, P. R. (1997). A didactic example of multilevel structural equation modeling applicable to the study of organizations. *Structural Equation Modeling* 4(1): 1-24.

Keats, B.W., Hitt, M. (1988). A causal model of linkages among environmental dimensions, macro organizational characteristics, and performance. *Academy of Management Journal* 31(3): 570-598.

Kraemer, H.C., Offord, D.R., Jensen, P. J. , Kazdin, A. E., Kessler, R.C., Kupfer, D. J. (1999). Measuring the potency of risk factors for clinical or policy significance. *Psychological Methods* 4(3): 257-271.

Lakka, T.A. et al. (1994). Relation of leisure-time physical activity and cardio-respiratory fitness to the risk of acute myocardial infarction in men. *New England Journal of Medicine* 330: 1549-1559.

Liebowitz, J. (ed.) (1999). *Knowledge Management: Handbook.* New York: CRC Press.

Lilienfeld, A.M. (1976). *Foundations of Epidemiology.* N.Y.: Oxford University Press.

Little, T.D., Schnabel, K. U., Baumert, J. (2000). *Modeling Longitudinal and Multilevel Data.* Mahwah, N.J.: Lawrence Erlbaum.Associates, Publishers.

Long, J. S. (1983). *Confirmatory Factor Analysis.* Beverly Hills, CA: Sage Publications, Inc.

Long, J. S. (1983). *Covariance Structure Models: An Introduction to LISREL.* Beverly Hills, CA: Sage Publications, Inc.

Luft, H. S. (1990). The applicability of the regression-discontinuity design in health services research. In L. Sechrest, E. Perrin, and J. Bunker (eds.), *Research Methodology: Strengthening Causal Interpretations of Nonexperimental Data.* Rockville, MD: Agency for Health Care Policy and Research.

Marks, R.G. (1982). *Designing a Research Project: The Basics of Biomedical Research Methodology.* Belmont, CA: Lifetime Learning Publications.

Maruyama, G. M. (1998*). Basics of Structural Equation Model.* Thousand Oaks, CA: Sage Publications, Inc.

Morgenstern, H. (1998). Ecologic studies. In K.J. Rothman and S. Greenland (eds.), Modern *Epidemiology.* Chestnut Hill, MA: Lippincott Williams & Wilkins, pp. 459-480.

Mulaik, S.A. (1987). Toward a conception of causality applicable to experimentation and causal modeling. *Child Development* 58: 18-32.

Murray, D.M. (1998). *Design and Analysis of Group-Randomized Trials.* N.Y.: Oxford University Press.

Muthén, B.O. (1994). Multilevel covariance structure analysis. *Sociological Research Methods and Research* 22(3): 376-399.

Muthén, B.O. (1991). Multilevel factor analysis of class and student achievement components. *Journal of Educational Measurement* 28(4): 338-354.

Norell, S.E. (1992). *A Short Course in Epidemiology.* N.Y.: Raven Press.

Ostrom, C.W. (1978). *Time-Series Analysis: Regression Technique.* Beverly Hills, CA: Sage Publications.

Phillips, K.A., et al. (1998). Understanding the context of health care utilization: Assessing environmental and provider-related variables in the behavioral model of utilization. *Health Services Research* 33(3): 571-596.

Polit, D.F., Hungler, B.P. (1987). *Nursing Research: Principles and Methods.* Philadelphia: J.B. Lippincott Company.

Reed, P. J. (1998). Medical Outcomes Study Short Form 36: Testing and cross-validating a second-order factorial structure for health system employees. *Health Services Research* 33(5): 1361-1380.

Rothman, K. J., Greenland, S. (1998). Causation and causal inference. In K. J. Rothman and S. Greenland (eds.), *Modern Epidemiology*. N.Y.: Lippincott Williams & Wilkins.

Rothman, K.J. (1988). *Causal Inference*. Chestnut Hill, MA: Epidemiology Resources, Inc.

Scott, S.G., Bruce, R.A. (1994). Determinants of innovative behavior. *Academy of Management Journal* 37(4): 580-607.

Short, L. M., Hennessy, M. (1994). Using structural equations to estimate effects of behavioral interventions. *Structural Equation Modeling* 1(1): 68-81.

Spector, P.E. (1981). *Research Design*. Beverly Hills, CA: Sage Publications.

Susser, M. (1973). *Causal Thinking in the Health Sciences*. New York: Oxford University Press.

Szeinbach, S. L., Barnes, J. H., Summers, K. H. (1995). Comparison of a behavioral model of physicians' drug product choice decision with pharmacists' product choice recommendations: A study of the choice for the treatment of panic disorder. *Structural Equation Modeling* 2 (3): 232-245.

Szklo, M., Nieto, E. J. (2000). Identifying noncausal associations: Confounding. In *Epidemiology: Beyond the Basics*. Gaithersburg, MD: Aspen Publishers, Inc.

Takkouche, B., Cadarso-Sqarez, C., Spiegelman, D. (1999). Evaluation of old and new tests of heterogeneity in epidemiologic meta analysis. *American Journal of Epidemiology* 150(2): 206-215.

Trochim, W.M.K. (1990). The regression-discontinuity design. In L. Sechrest, E. Perrin, and J. Bunker (eds.), *Research Methodology: Strengthening causal interpretations of nonexperimental data*. Rockville, MD: Agency for Health Care Policy and Research.

Wan, T.T.H. (1992). Hospital variations in adverse patient outcomes. *Quality Assurance and Utilization Review* 7: 50-53.

Wan, T.T.H. (1995). *Analysis and Evaluation of Health Systems: An Integrated Approach to Managerial Decision Making*. Baltimore: Health Professions Press.

Wan, T.T.H., Pai, C.W, Wan, G. J. (1998). Organizational and market determinants of HMOs' performance in preventive practice. *Journal of Health Care Quality* 20(3): 14-129.

Wan, T.T.H. (2001). Assessing causality: Foundations for population-based health care managerial decision making. In Denise M. Oleske (ed.) *Epidemiology and the Delivery of Health Care Services: Methods and Applications*, 2nd edition. New York: Plenum Press/Kluwer Academic Publishers.

Weed, Douglas L. Causal criteria and Popperian refutation. In Kenneth J. Rothman (ed.), *Causal Inference*. Chestnut Hill, Massachusetts: Epidemiology Resources, Inc., 1988.

CHAPTER 3

RESEARCH ON HEALTH SERVICES MANAGEMENT: THE SEARCH FOR STRUCTURE

Health services management research (HSMR) has been described as an iterative process, which compiles data, generates pertinent managerial information, and influences the delivery of health care services. This field of inquiry examines the effects of organization, financing, management, and market forces on the performance of health care organizations in the delivery, quality and cost of, access to, and outcomes of health services. The inquiry encompasses a variety of disciplines: social and behavioral sciences, statistics, economics, sociology, political science, anthropology, psychology, operations research, epidemiology, and biostatistics, as well as medicine and nursing. Responsibilities of HSMR include:

1. Collection and diffusion of health care information and statistics.
2. Design, development, and evaluation of new health services systems and processes.
3. Improvement of health services through quality assessment, improvement and management.
4. Research and theory building in management.
5. Policy impact analysis.

CONTRIBUTIONS OF HEALTH SERVICES MANAGEMENT RESEARCH

The contributions of HSMR can be viewed in terms of its research foci and historical development. Mechanic (1978), in a review of the foci of health care research during 1970-1980, reports that one of the most acute needs of administrators is to know the facts about gaps in the distribution of services; actual costs of medical and surgical procedures in different localities; relationships between expenditures and changes in health status; rates of admission to hospitals and lengths of stay for varying procedures, and ways they are changing; costs of new technologies and how they affect physician behavior and medical outcomes; and many other aspects of hospital management.

Greenlick, Freeborn and Pope, (1988) summarize the contributions of research activities during 1980-1990 as follows:

"Medical care utilization remains a complex problem. The ability to explain interpersonal differences in such utilization, even within a single medical care system, has not advanced very far. Also, assessing the impact of organization of care on utilization remains a basic research issue. As the utilization model gets more complex and more specific, the need for systematization grows. The thread of research from the early posthospital studies and the home care--extended care facility study--through the studies of utilization of the aged and the Medicare demonstration projects will help create the knowledge base for organizing the medical care system in the next century, when more than 20% of the population will be over 65. The work in health behavior methodology and in randomized trials for health care alternatives will continue to grow. As more profit-making corporations enter the field, market forces become increasingly competitive and health care becomes more market-driven than mission-driven. Where this trend will lead, whether profit-making institutions will be able to meet the functional requirements of the medical care system without distorting its social purpose, and how all of this will affect the integrity of research in the public domain, are questions that loom large in considering the future of the American health care system."

Between 1990 and 2000, the application of managerial epidemiology to the delivery of health care services has increased understanding of the distribution of the needs for health care, including disease, impairments, disability, injuries, and other health problems in human populations, and also of the factors contributing to their emergence, severity and consequences (Oleske, 2001). The use of the contingency strategic adaptation framework has identified causal factors that are amenable to change or control so that managerial interventions can achieve better patient outcomes and organizational performance (Wan, 1995). A new research paradigm has evolved which investigates the effects of both individual and aggregate (community or organizational) explanatory factors on the use of health services and on patient outcomes. For example, Andersen (1995) proposed a revised model of health behavior that simultaneously considers individual and ecological factors in health behavior and health care outcomes (Figure 6). Another prominent in health care management research is the population-based approach to the study of health service delivery systems (Wan, 1995). The use of a comprehensive framework developed by Aday, Begley, Lairson, and Slater (1993) has helped to identify family, community, and organizational factors that will improve the delivery of health care services and thus will improve the quality of life or well being of the population. This approach has broadened the scope of outcomes research, which studies the end results of medical care or the effects of health care on the health and well being of patients and the population. Many management studies fall into one of these two categories: 1) Efficacy studies examine the success of treatments in controlled environments; and 2) effectiveness studies examine real-life settings (the doctor's office, hospital, clinic, or home). Sponsored programs/research have identified the variations in medical care (Ginzberg, 1991; Wennberg, 1984), compared the effectiveness of various treatments and procedures (PORT Studies funded by the Agency for Health care Research and Quality), developed appropriate criteria for evaluating managed care organizations (e.g., the National Committee for Quality Assurance's initiatives), and measured hospital performance (Ozcan et. al., 1998; Wan, 1995) and patient satisfaction (Kersnik, 2001; Ware et al., 1995). The results of outcomes research have been disseminated to health professionals with the aim of changing the behavior of providers, patients, health care institutions, and payers. Consequently, outcomes research has prompted: (1) clinical practice guidelines, (2) changes in the accreditation process, (3) new reimbursement policies, and (4) quality improvement/management initiatives.

Over the past three decades, health care research has also generated scientific information to guide policy analysis and evaluation (Bice, 1980; Shortell and Kaluzny, 2000; White, 1992;). For example, policy analysis examines cost containment (Klarman, 1980), health status (Ware, 1986), professional behavior and practice (Gray, 1997), and forces affecting the health of the population (Andersen, 1995). Analysts evaluate past and existing financial arrangements and documentation of emerging health care problems (Ginsberg, 1997). Official annual publications now document the determinants of equitable distribution and use of health services and resources (Public Health Service, 1999).

In the new era (2000+), information technology has the potential to dramatically sharpen the focus of the health care system on service needs and also on preferences (Kendall and Levine, 1998; Institute of Medicine, 2001). Effective managed care organizations rely on the availability of accurate electronic data on administrative, financial, market, and clinical information. Clearly, it will be necessary to systematically integrate information with what is known about health care management in order to establish evidence-based management. Thus, knowledge management is a new field of investigation in health services management. Readers can have access to a variety of references by clicking www.brint.com or www. kmnews.com. Information on market forces, provider and organizational systems, and patient and population characteristics is particularly germane in understanding the variation in health organizations' performance. Because the unit of analysis varies by the type of data the investigator collects, it is imperative to understand how factors at both the individual (patient) and the organizational levels contribute to the variations in organizational performance and in patient outcomes. Therefore it is critical to apply multilevel analysis to the complex data set, so that the between-group and within-group variations can be accounted for (Heck and Thomas, 2000). There are ample opportunities for innovative studies that will contribute to society. It is anticipated that collaboration between researchers and corporate executives in the health industry will further the understanding of how to do the right things (e.g., improve quality as well as efficiency) and how to do so effectively (attain high performance).

THE SEARCH FOR THE STRUCTURE: FROM INFORMATION MANAGEMENT TO KNOWLEDGE MANAGEMENT IN HEALTH ORGANIZATION RESEARCH

The explicit and implicit knowledge in organizational research can be structured under an integrated framework of health systems analysis that specifies the relationships among variables of organizational context, design, performance and outcome (Figure 6). The causal specifications of this framework assume that linear relationships exist among the four domains of the variables, and that the temporal sequences of those variables follow an order of context-design-performance-outcome. The indicators or measurement variables associated with each dimension of the framework can be classified into a specific domain. In each domain, variables can be further divided into latent theoretical constructs and observed indicators. The selection of study variables is based on the literature in health services management research.

CONTEXT $\rightarrow$	DESIGN $\rightarrow$	PERFORMANCE $\rightarrow$	OUTCOME
Market competition	Integration	Efficiency	Adverse or sentinel
Managed care penetration	strategies	Effectiveness	health events
Location	Care modalities	Productivity	Functional outcomes
SES & resources	Management		Patient satisfaction
Demographics	interventions		

Figure 6. Framework for Evaluating Health Care System Performance

Figure 7 shows how study variables can be structured so that a data warehouse for health services management research can be built. A data warehouse is defined as a collection of decision support functionalities to enhance executives or managers to make systematic and relevant decisions

(Kerkri, et al., 2001). Multiple datasets will be compiled with coherent and consistent definitions of variables or constructs in building an integrated data repository. The data warehouse is a subject-oriented and relational database. The temporal specification of the variables is an important feature of the data system. Longitudinal data should be compiled for trend analysis and causal inference.

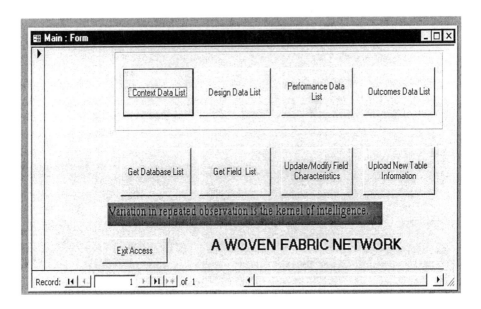

Figure 7. A Data Warehouse for Health Systems Analysis

An integrated system of health information is greatly needed in health services management. In building such a system, three major functions of a wrapper are translation, monitoring, and anonymity (Kerkri et al., 2001). The translation is to collect multiple sources of clinical, administrative, and managerial data within the same organization and to convert data from different formats into a common format. Any changes in a data system should be detected and monitored so the stability or consistency of data is maintained and enhanced. Using an encryption algorithm to ensure the security and confidentiality should protect data gathered. Data conflicts or inconsistencies should be resolved before the data are populated in a data warehouse.

Figure 8 shows an example of an enterprise-wide data systems in health informatics. Information generated can be used for monitoring and evaluating organizational performance and patient care outcomes. Ultimately, expert systems should be developed to guide executive and clinical decision-making.

Figure 8. Health care Information for Management

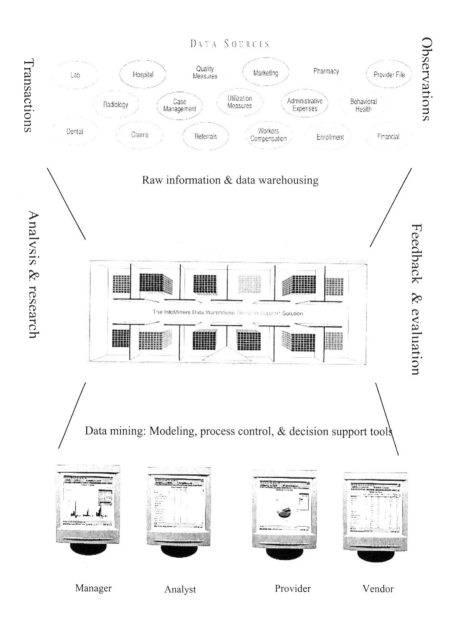

Several data software packages can be used to establish a relational database, including SPSS, SAS, and SiliconGraphics. The capabilities of these software packages vary slightly.

Irrespective of the type of software used, the fundamental principles of building a successful data warehousing project are as follows: 1) Use a clearly specified theoretical framework. Shortell and Kaluzny (2000) suggest several theoretical frameworks to guide the design and analysis of health care organizations. 2) Formulate a data dictionary with operational definitions for each variable and documentation of the data sources. 3) Clarify measurement issues pertaining to the study variables and their statistical distributions. 4) Select pertinent and quantifiable indicators to measure the constructs. 5) Identify temporal sequences of the study variables. 6) Validate correlated variables to determine if there are potential collinearity problems. 7) Use a common identifier to merge multiple datasets properly and accurately.

The data warehouse is structured with relational data sets from primary sources such as the clinical and administrative units of health care organizations, and secondary sources such as the American Hospital Association's annual surveys, JCAHO's accreditation reports, governmental reports, and community surveys. Data compiled from internal operations should be updated constantly to form a core information base to support managerial and decisions. Ideally, clinical and administrative data should be integrated into an enterprise data system (Figure 8). Information from both internal and external sources should be structured and amended easily when new data become available. For instance, if a data warehouse is built for strategic planning and assessment of service capacities, data for each study unit should be compiled. The more comprehensively the data warehouse is structured, the better the information is for data mining and exploratory functions. Ultimately, the databases should provide critical information and generate pertinent knowledge for improving a system's performance.

Data Mining: Exploratory and Confirmatory Approaches

Exploratory Analysis

Research in health services management seeks to explore the data to discover characteristic features and interesting relationships without imposing any definite model on the data. Exploratory statistical analysis (ESA) gives researchers various possibilities. Researchers should be open to possibilities that they do not expect to find, especially in the case of weak theories that specify no models for the relationships between variables (Roberts and

Burke, 1989). Exploratory analysis may suggest structural relationships among multiple variables, formulate plausible hypotheses for testing, and build alternative models for validation.   Such a knowledge management process assembles unstructured data and explores them to generate new knowledge. The statistical techniques or analytical methods for exploratory analysis include: 1) component analysis, 2) factor analysis, 3) cluster analysis, 4) multidimensional scaling, 5) descriptive statistical measures, 6) graphical display, and 7) automatic interaction detector analysis (AID). These analytical methods will be presented in the next chapter.

Confirmatory Approach

An investigator relies on a theoretical framework in building a specific model to describe, explain or account for the data in terms of relatively few parameters. The model is based on a priori information about the data structure (theory or hypothesis, design, knowledge from previous research). On the basis of available structured data, the investigator wants to validate the model and to test hypotheses about the parameters of the model.   In addition, the goodness of fit statistics will be used to evaluate the model fit. When a model is not supported by the data, alternative models can be formed and examined. This confirmatory approach to knowledge management is an iteration of processes: model designing, hypothesis testing, trimming, and validating (Figure 9).

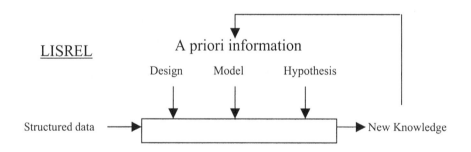

Figure 9. Iterative Process in Generating New Knowledge
Source: Jöreskog, K.G. and Sörbom, D. LISREL: Lecture Notes, 1981.

The statistical method used in the confirmatory approach is based on Linear Structural Relationships (LISREL) modeling. For a program evaluation on hospital performance, researchers often need to analyze several types of outcomes (e.g., the complication rate, repeated hospitalization rate, hospital mortality rate, etc.). Sometimes these outcome variables are correlated with each other. In that case, use of a multivariate statistical technique is essential to examine the effect of an intervention on multiple outcome variables, with or without correlated errors or residuals. The outcome variables are treated as endogenous variables, and the intervention variable is treated as an exogenous variable.

LISREL analysis of quantitative outcome variables has been demonstrated for its usefulness in data analysis and theory construction (Pai and Wan, 1997; Reed, 1998; Wan, 2001). The LISREL model contains two parts. The *measurement model* specifies how the theoretical constructs or latent variables (e.g., adverse health care outcomes) are measured by observable indicators (e.g., hospital mortality rate, complication rate, etc.). The other part is the *structural equation model*, which represents the causal relationships among the exogenous and endogenous variables. The structural equation model is like the path-analytic model in three aspects: (1) model construction, (2) parameter estimation of the model, and (3) testing the fit of the model to the data by comparing observed correlations with predicted correlations among the study variables. However, the LISREL model is less restrictive than the path-analytic model. For instance, it allows the investigator to ask direct questions about the data, in the form of different restrictions on the coefficients. LISREL can easily handle errors in measurement, correlated errors and residuals, and reciprocal causation. Thus, LISREL's advanced procedures can specify, compare, and evaluate the effects of an intervention or regulatory changes on a set of correlated outcome variables, using latent growth curve modeling techniques (Porter, 2001).

In assessing the outcomes of "outcomes and effectiveness research" funded by the Agency for Health care Research and Quality, Stryer and associates (2000) conclude that the challenges for the next-generation health services research should be to advance from hypothesis generation to definitive studies of effectiveness. Furthermore, outcomes research should provide definitive answers to "what works" to improve health policy, practice, and health care outcomes. This conclusion signifies an important direction in future studies to generate pertinent information for improving evidence-based health care management and practice.

Researchers on health services management should also consider the relationships among different outcomes. Failure to do so can lead to inappropriate conclusions about program effects. Structural equation modeling not only allows assessment of the effects of the intervention on multiple outcome (endogenous) variables, but also allows assessment of the net effect of the intervention variable when the effects of other factors are simultaneously controlled.

A variety of epidemiological statistical techniques are appropriate in longitudinal program evaluations. Structural equation modeling, applied to the panel data described earlier, is a useful method of panel analysis. Because this technique analyzes the relationships between endogenous (outcome) variables, one can examine the relationship between those variables across time as well. For example, structural equation modeling can control for other factors when assessing how the use of various types of community-based long-term care is related to later use of nursing home services.

Meta Analysis

This is a popular analytical strategy for assessing program outcomes, which can be used to estimate the effect size (ES), or intervention effect, of specific program outcomes. Properly used, meta analysis can strengthen causal interpretations of non-experimental data. Furthermore, evaluation researchers can develop a covariance structural model for multiple, comparable studies and then empirically test the model's goodness of fit by using meta analysis. For example, the effect of case-managed services on such geriatric patient outcomes as quality of life (a latent construct) can be evaluated by pooling different study samples together if comparable outcome measures are used. In that case, the measurement model of quality of life is first evaluated; then the equality constraints for observed indicators are assumed for the multiple samples. The net effect of case-managed services on patient outcomes in the study groups is thus measured while the effect of other extraneous factors is simultaneously controlled.

CONCLUSION

The challenge for health services management and education in the 21st century is to apply scientific methods and confirmatory analyses to specific managerial problems and decision-making areas. To begin this process of scientific inquiry, the following are suggestions for health care management researchers:

1. Health Information and Technology: Understanding the emerging contributions of informatics, decision support systems, and expert systems to redesign or restructure the health care delivery system.
2. Knowledge Management: Maximizing the productivity or performance of health care organizations by employing knowledge management strategies and tactics.
3. Analytical Skill Development: Using sophisticated tools and techniques available for data mining and simulation.

4. Meta Data for Developing Executive Decision Support Systems: Investing resources to build a meta data repository and to integrate clinical, administrative, and managerial data for clinical and managerial decisions.
5. Collaboration between Academic and Practical Fields: Establishing strong research collaboration between the health administration faculty and practitioners.

In conclusion: concerted efforts should be made to employ data warehousing and data mining strategies in health care management and research. Scientifically derived evidence should be gathered to guide the integration of care services, to improve the quality, equity, and efficiency of health care.

REFERENCES

Aday, L., Begley, C.E., Lairson, D.R., Slater, C.H. (1993). *Evaluating the Medical Care System: Effectiveness, Efficiency, and Equity.* Ann Arbor, Michigan: Health Administration Press.

Andersen, R.M. (1995). Revisiting the behavioral model and access to medical care: Does it matter? *Journal of Health and Social Behavior* 36(March): 1-10.

Bice, T.W. (1980). Social health services research: Contributions to public policy. *Milbank Memorial Fund Quarterly* 58(2): 173-200.

Ginzberg, E. (1991). *Health Services Research: Key to Health Policy.* Cambridge: Harvard University Press.

Ginzberg, P. (1997). A perspective on health system change in 1997. *Charting Change: A Longitudinal Look at the American Health System, 1997 Annual Report.* Washington, D.C.: Center for Studying Health System Change.

Gray, J.A.M. (1997). *Evidence-Based Health Care: How to Make Health Policy and Management Decisions.* N.Y.: Churchill Livingstone.

Greenlick, M.R., Freeborn, D. K., Pope, C. R. (1988). *Health Care Research in HMO: Two Decades of Discovery.* Baltimore: The Johns Hopkins University Press.

Hech, T., Thomas, S.L. (2000). *An Introduction to Multilevel Modeling Techniques.* Mahwah, NJ: Lawrence Erlbaum Associates, Publishers.

Institute of Medicine. (2001). *Crossing the Quality Chasm: A New Health System for the 21st Century.* Washington, D.C.: National Academy Press.

Jöreskog, K.G. Sörbom, D. LISREL: Unpublished Lecture Notes, 1981.

Kendall, D.B., Levine, S.R. (1998). Pursuing the promise of an information-age health care system. *Health Affairs* 17(6): 41-43.

Kerkri, E.M., Quantin, C., Allaert, E.A., Cottin, Y., Charve, Ph., Jouanot, F., Yétongnon, K. (2001). An approach for integrating heterogeneous information sources in a medical data warehouse. *Journal of Medical Systems* 25(3): 167-176.

Kersnik, J. (2001). Determinants of customer satisfaction with the health care system. *Health Policy* 57(2): 155-164.

Klarman, H.T. (1980). Observations on health services research and health policy analysis. *Milbank Memorial Fund Quarterly* 2(2): 201-216.

Mechanic, D. (1978). Prospects and problems in health services research. *Milbank Memorial Fund Quarterly* 56(2): 127-139.

Oleske, D. M. (ed.) (2001). *Epidemiology and the Delivery of Health Care Services: Methods and Applications.* 2nd edition. New York: Kluwer Academic/Plenum Publishers.

Ozcan, Y. (1995). Efficiency of hospital service production in local markets: The balance sheet of U.S. medical armament. *Socio-Economic Planning Sciences*, 29 (2), 139-150.

Pa, C.W., Wan, T.T.H. (1997). Confirmatory analysis of health outcome indicators: the 36-item short-form health survey (SF-36). *Journal of Rehabilitation Outcomes Measurement* 1(2): 48-59.

Porter, S.J. (2001). A longitudinal analysis of the distinction between for-profit and non-for-profit hospitals in America. *Journal of Health and Social Behavior* 42:17-44.

Reed, P.L. (1998). Medical outcomes study short form 36: Testing and cross-validating a second-order factorial structure for health system employees. *Health Services Research* 33(5): 1361-1380.

Roberts, C.A., Burke, S.O. (1989). Nursing Research: A Quantitative and Qualitative Approach. Boston: Jones and Bartlett Publishers.

Shortell, S.M., Kaluzny, A.D. (2000). *Health Care Management: Organization, Design, and Behavior.* 4th edition. Albany, NY: Delmar Publishers.

Stryer, D., Tunis, S., Hubbard, H, Clancy, C. (2000).   The Outcomes of Outcomes and Effectiveness Research: Impacts and Lessons from the first decade. *Health Services Research* 35(5): 977-993.

Wan, T.T.H. (1995).   *Analysis and Evaluation of Health Care Systems: An Integrated Managerial Decision-Making Approach*. Baltimore: Health Professions Press.

Ware, J. (1986). The assessment of health status. In L. Aiken and D. Mechanic (eds.) *Applications of Social Science to Clinical Medicine and Health Policy*. New Brunswick, NJ: Rutgers University Press.

Wennberg, J. (1984). Dealing with medical practice variations: A proposal for action. *Health Affairs* 3: 6-32.

White, K.L. (1992). *Health Services Research: An Anthology*. Washington, D.C: Pan American Health Organization.

CHAPTER 4

EXPLORATORY ANALYTICAL MODELING METHODS

In analyzing and testing unexplored areas of health services management research in which theories have not yet matured, exploratory statistical analysis (ESA) gives researchers a wide variety of possibilities. Researchers should be open to unforeseen discoveries, especially in the case of weak theories that specify only that variables are related to each other (Hartwig and Dearing, 1979).

This chapter introduces certain ESA techniques that are useful in formulating the underlying structure of data:

1. Automatic interaction detector (AID) analysis: principles and application
2. Logistic regression analysis: principles and application
3. Path analysis: principles and application
4. Factor analysis: principles and application.

AUTOMATIC INTERACTION DETECTOR (AID) ANALYSIS

Principles

Sonquist, Baker, and Morgan (1975) developed an analytical strategy to search for the best model to explain the main effects and interaction effects of multiple independent variables on a dependent variable. AID can be used in the exploratory stage of model building and is applicable to sample survey data in which you have:

1. A set of observations or units of analysis.
2. A set of predictors $X_1, X_2, \ldots X_p$, measured on a nominal or ordinal scale.
3. A dependent variable Y, measured on an interval or ratio scale.
4. Data sets appropriate for AID include those with a large number of cases (i.e. a thousand or more), a well-behaved dependent

variable without extreme cases or severe bimodalities, dichotomous variables, and categorical predictor variables.

A non-technical description of AID is as follows:

- AID subdivides the original sample, through a series of dichotomous (binary) splits with regard to the predictors, into a number of mutually exclusive subgroups.
- Every observation becomes a member of exactly one of these subgroups.
- Iterative procedure generates the terminal subgroups.

The steps of AID are:

First, from all potential dichotomies, the predictor and that division of categories of the predictor that will maximize between-group variance are chosen.

Second, the subgroup with the largest sums of squares with regard to the dependent variable is divided to maximize the explained variance. Analysis continues with successive splits until the minimum requirements for size and variance are no longer met by any of the subgroups of the sample (Andersen, Smedby, and Anderson, 1970).

In the AID analysis, a dependent variable (Y) is studied in relation to a series of independent variables or predictors $(X_1, X_2 \ldots)$. The software program subdivides the original sample, through a series of dichotomous splits with respect to the predictors, into a number of mutually exclusive subgroups. This is achieved through a stepwise procedure. In the first step, the predictor, and the split of that predictor, are chosen that explain as much as possible of the total sampling variance. Through this first split the original sample is divided into two subgroups. In the second step, starting with the subgroup with the greatest variance, the program chooses that predictor and that split that explains as much as possible of the remaining variance. The analysis continues with successive splits so that in each step, a split is made which explains as much as possible of the remaining variance of the parent group. The requirements that regulate the partition process are as follows:

Split Eligibility Criterion. It specifies the proportion of the total sums of squares that must be contained in any group if that group is to become a candidate for splitting. The requirement is made to prevent groups with little variation from being split.

Minimum Group Size Criterion. It prevents groups with small numbers of observations from splitting. The size decision is very arbitrary and depends upon an investigator's research interest.

Split Reducibility Criterion. It specifies what minimum proportion of the total sum of squares must be explained by the best possible split of a group in order to allow that group to be split. If this requirement is not met, it means that there is no useful predictor. Thus, $Q <$ or $=$ BSS$_i$, where BSS refers to the between sum of squares for group i, predictor x_k, and partition p of this group into two non-overlapping subgroups.

Theoretical Import Of Independent Variables. Predictors can be ordered in terms of their relevancy to the dependent variable so that predictors can be entered sequentially in the AID analysis. For instance, use of health services can be explained by three groupings of predictors: 1) predisposing factors— factors that predispose an individual in his or her use of health services; 2) enabling factors that impede or facilitate the use of health services; and 3) need for care factors that influence or trigger the health or illness behavior (Figure 10). Detailed definitions of dimensions can be found in several publications (Andersen and Newman, 1973; Wan and Soifer, 1974; Wan and Yates, 1975).

Determinants

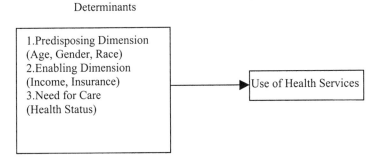

Figure 10. Determinants of Health Services Use

Computation

Gross B_{K^2} = the maximum proportion of variance that could be explained by predictor x_k by one split of the first group (i=1) into two subgroups:

$$\text{Gross } B_{K^2} = \frac{BSS}{TSS}i.1$$

Partial B_{K^2} = the actual proportion of variance explained by predictor X_k in the whole analysis:

$$\text{Partial } B_{K^2} = \frac{\sum TSS_{ik} - \sum TSS_{jk}}{TSS_T},$$

where ΣTSS_{ik} is the sum of squares over all parent groups actually split by predictor X_k, and ΣTSS_{jk} is the sum of squares over all new groups formed by splitting a parent group on predictor X_k.

The total proportion of variance explained by the analysis is defined as the sum of the partial B^2's where the summation is extended over all predictors actually used in the analysis.

Predictor Tree: A diagram presents the parent groups and their subgroups.

The following algorithm uses the binary split approach to partition a set of data:

1. The independent variables and their categories are ordered in accordance with their relevance to the dependent variable.
2. A whole data set (total sample) is treated as the beginning parent group at the initial stage.
3. Based upon the principle of one-way analysis (reduction of the total variance TSS), subgroups are formed by splitting the parent group into mutually exclusive subgroups. Therefore, the within-group difference with respect to the dependent variable is minimized, and the between-group difference is maximized by the predictor.

A Windows-based software, CART developed by Salford Systems (Steinberg and Colla, 1995), can be used to generate the predictor tree and AID results.

Two-Stage AID Analysis

The AID program allows a two-stage analysis. After the first stage, which includes an AID analysis, residuals are calculated for each observation in the sample and saved for subsequent analysis. The residual is the difference between the observed and predicted values of the dependent variable and the predicted value. The predicted value of the dependent variable is the mean of the final group of which the observation is a member. In the second stage of AID analysis, a new AID analysis is performed using the residual of the dependent variable in the first stage as the new dependent variable. For example, in the first stage of AID analysis, the dependent variable, number of physician visits, can be analyzed by demographic and social economic predictor variables. Then the residual of the first-stage analysis can be used as the dependent variable in the second-stage analysis, with the need for care factors as the predictor variables.

Application of AID Analysis and Methodological Issues

Health Applications

AID analysis has been applied to the management of health services in several key areas: diagnostic-related groupings (DRGs) applied in the prospective payment system (Fetter, et al., 1989); all-payer-based DRGs (AP-DRGs); the ambulatory patient classification system (www.hcfa.gov/medicare); resource utilization groupings (Fries, 1990); service planning groupings (Wan and Yates, 1974); and case mix groups for the prospective payment system used in inpatient rehabilitation facilities (www.hcfa.gov/medicare/irfpps.htm). In these applications, systematic ways of unraveling complex variable dependencies and statistical interactions were examined to establish prediction equations explaining a large part of the variation in the dependent variables (Wan and Soifer, 1974; Wan, Weissert, and Livieratos, 1980).

The major AID applications to health services management are exemplified in the current development of a prospective payment system for inpatient rehabilitation facilities. This system consists of three parts (Health Care Financing Administration, 2000). 1) Using a patient classification system, mutually exclusive groups of patients, based on similar clinical and functional characteristics and similar levels of resource use, are generated to form distinctive case-mix groups. The functional independence measures used are age, impairment scores, cognitive functioning scores, and other measures of functional capability. Data generated from 624 inpatient rehabilitation facilities in 1996-1997 are used to create the CMG classifications. 2) Relative weights are derived that are proportional to the resources needed by a typical member of a case mix group. 3) In setting the payment rates, these cost adjustments factors are included: local wages, case levels in outliers, transfers and interrupted stay cases, disproportional share of severe patients, and rural-urban location.

Methodological Issues of AID Analysis

The AID analysis is limited for several reasons. First, the statistical stability of the predictor tree must be validated using randomly selected sub-samples of the original sample. Second, size restrictions complicate the problem of producing logical and substantively meaningful splits of groups in the predictor tree. The requirement of a minimum size for each group can inhibit the proper use of AID analysis. Third, causal inference is not attainable, because temporal sequences of the factors involved often are unspecified in the cross-sectional study design, thus making it difficult to delineate causal factors for performance or outcome. Fourth, exploratory models must be verified and tested to generate consistent and logical interpretations of the observed evidence. Fifth, the proposed exploratory model should be replicated with a different sample.

LOGISTIC REGRESSION ANALYSIS

The work of Cornfield and Kannel (1967) in which they used logistic regression for multivariate analysis of the Framingham Heart study exemplifies a logistic regression model in its full strength and shows the applicable power of this analysis. The method became the method of choice in health sciences, because most of the explanatory variables in this field are dichotomous.

The logistic regression model is used when the dependent variable is measured by a binary or discrete variable, and the independent variables (risk factors or interventions) are continuous and discrete. Since the dependent variable is a discrete variable (e.g., the probability of being hospitalized in a specified period), the predicted probability should lie in the unity boundary. Logistic regression is preferable to ordinary least squares (OLS), because OLS estimates are biased and yield predicted values that are not between 0 and 1. If p is the probability, we assume that the logit $(p) = \dfrac{p}{(1-p)}$ is a linear function of the predictor variables, or in other terms, that $p(x) = \dfrac{1}{(1 + e^{-bx})}$. When the probability is relatively small, p(x) is roughly equivalent to e^{bx}.

The logistic model is expressed either in terms of the log-odds (the ratio of two individual odds) for a given outcome (e.g., improved population health) or in terms of the probability of that outcome (e.g., the probability of improved population health). The log-odds is assumed to be a linear function of the magnitude of intervention instituted by a program, or log-odds = l_x = a + bx, where l_x represents the logarithm of the odds of improved health for a specific continuous value x of the intervention/treatment variable. The coefficient b measures the change (multiplicative) in the likelihood of having improved health that is associated with a one-unit change in the intervention variable on the log-odds scale; e^b measures the change in population health associated with a one-unit change in the intervention variable on the odds scale.

Table 3 is an example of using multiple logistic regression analysis to predict a hospital's mortality risk for patients with abdominal aortic aneurysms. Two models were developed to identify the mortality risk: Model 1 includes only demographic factors and admission status; and Model 2

52

includes demographic factors, admission status, and a number of complications. Table 3 reports the estimated coefficient (B), estimated standard error (SE), and odds ratio (relative risk). The odds ratio, which refers to the relative risk of dying from an abdominal aortic aneurysm, is computed by exponentiating the estimated coefficient for a given predictor. In Model 1 with only three predictor variables (age, gender and admission status), patients 70 years and older were 4.29 times more likely to die than those younger than 70. In Model 2, additional predictor variables were added to reflect the clinical status of patients. A comparative statistic for the difference between two log-likelihood ratios for the two models is 24.06 (96.68-72.62) with 4 degrees of freedom. Evidently, Model 2, with a smaller log-likelihood ratio, is a better model in accounting for the mortality differentials.

The reader who wants to learn more about the application of multiple logistic regression should read the books by Hosmer and Lemeshow (1989) and Maddala (1983).

Table 3. **Multiple Logistic Regression Analysis of Mortality Risk for Abdominal Aortic Aneurysm (N = 174 Patients Treated): Two Models**

Variable	Model 1				Model 2			
	B	SE	B/SE	Odds Ratio	B	SE	B/SE	Odds Ratio
Age (70+=1; <70=0)	1.46	0.62	2.35*	4.29	1.58	0.81	1.95	4.84
Gender (Female=1; Male=0)	1.24	0.56	2.21*	3.47	1.65	0.68	2.43*	5.19
Admission Status (Elective=0; Non-elective=1)	2.32	0.59	3.93*	10.16	2.24	0.70	3.20*	9.36
Ventricular Tachycardia (Yes=1; No=0)					2.24	1.05	2.13*	7.98
Peri-OP Myocardiac Infarction (Yes=1; No=0)					2.52	1.07	2.36*	12.37
Acute Renal Failure with Dialysis (Yes=1; No=0)					2.17	0.98	2.21*	8.73
Pneumonia (Yes=1; No=0)					1.71	0.80	2.14*	5.53
Constant (intercept)	-4.04	0.64			-5.25	0.93		
Log-likelihood	-96.68				-72.62			
(degrees of freedom)	(3)				(7)			

*Significant at 0.05 or lower

FACTOR ANALYSIS

When collecting information about unexplored complex issues in health services, researchers face the difficulty of studying large numbers of variables. In essence, the correlations to be considered reach huge numbers that are infeasible to examine. Thus, researchers must rely on data reduction techniques: factor analysis and principal component analysis. These techniques can systematically summarize large correlation matrices and uncover the underlying theoretical structure of the data.

Principal Component Analysis

Principal component analysis is a statistical technique that linearly transforms an original set of variables into a smaller set of uncorrelated variables. The purpose of principal component analysis is to extract factors (i.e., principal components) in order to explain most of the total variance in the data with the least number of factors. The principal components are extracted so that the first principal component, labeled as $PC_{(1)}$, explains the largest amount of the total variance.

$PC_{(1)}$ is the linear combination of the observed variables X_j, $j = 1,2,....p$, and $PC_{(1)} = a_{(1)1}X_1 + a_{(1)2}X_2 + ...a_{(1)p}X_p$ .... (4-1),

where $PC_{(1)}$ is the principal component with the largest variation; and $a_{(1)1}, {}_1a_{(1)2}, a_{(1)p}$ are the weights that have been chosen to maximize the ratio of the variance of $PC_{(1)}$. The second principal component, $PC_{(2)}$, is the weighted linear combination of the observed variables, which is uncorrelated with the first linear combination and accounts for the maximum amount of the remaining total variation not explained by $PC_{(1)}$ (Dillon and Goldstein, 1984, p. 25). In essence, the principal component is the weighted linear combination of the X's that accounts for the largest variance of all linear combinations that are uncorrelated with all of the previously extracted principal components. The general equation is:

$PC_{(m)} = a_{(m)1}X_1 + a_{(m)2}X_2 + ...w_{(m)p}X_p$ ....(4-2).

Principal component analysis can be derived by using the characteristic equation. Solving the equation (using the matrix notation) would produce values and vectors that are associated with a matrix. The equation can be written as: $XA = \lambda A$ .... (4-3),

where X is the matrix for which a solution is sought; A is the vector to be found; and λ is the value or a matrix of corresponding latent roots ordered from largest to smallest. The matrix can be solved to get the value of λ and a, the basic statistics of principal component analysis. The largest latent root of R is the variance of the first principal component analysis of R and its associated vector. The set of weights for the first principal component that maximize the variance can be written as:

54

$$a_1 = \begin{vmatrix} a_{11} \\ a_{12} \\ a_{13} \\ a_{14} \\ a_{15} \end{vmatrix}$$

Exploratory Factor Analysis

Factor analysis is another data reduction technique, which investigates interrelationships among variables to generate a new set of variables that express what is common among the original set. Factor analysis differs from principal component analysis, which considers the total variation contained in a set of variables. Factor analysis tries to identify that part of the common variance that is shared by the study variables. The aim of factor analysis is to simplify complicated and diverse relationships among variables by revealing common factors that link seemingly unrelated variables. Thus factor analysis help to reveal the underlying structure of data.

There are two types of factor analysis. The first is the exploratory factor model, which investigates the linkages between observed variables and unknown latent variables. The second is the confirmatory factor model, designed to test the hypothesized link between the observed variables and known underlying factors. On one hand, if the researcher has no theoretical hypotheses to guide him/her, the context is exploratory factor analysis. On the other hand, if the researcher has theoretical information on the structure of the data and wants to test that hypothesized structure, the context will be confirmatory factor analysis. Confirmatory factor analysis is discussed in greater detail in the following chapters.

Principles of Exploratory Factor Analysis

The basic common factor model can be written as

$X = \Lambda f + e$ ....(4-4),

where:

X is a p-dimensional vector of observed responses;

f is a q-dimensional vector of unobservable variables called common factors;

e is a p-dimensional vector of unobservable variables called unique factors; and

Λ is a p × q matrix of unknown constants called factor loadings.

We assume that there are p unique factors, and that the unique part of each variable is uncorrelated with each other's or with their common part; that is, Cov $(e,f) = 0$.

$$E(ee') = \Psi = \begin{vmatrix} \psi_1 & & & & \\ & \psi_2 & & & \\ & & .. & & \\ & & & .. & \\ & & & & \psi_p \end{vmatrix}$$

The model given by equation 4-4 along with the imposed assumptions imply that the covariance matrix of the response vector X, denoted by Σ_{xx}, can be expressed as

$$\Sigma_{xx} = \Lambda \, \Phi \, \Lambda' + \Psi \quad \text{.......................} \quad (4\text{-}5),$$

where Λ and Ψ are previously defined and

$$\Phi = \begin{vmatrix} 1 & & & & \\ \phi_{21} & 1 & & & \\ \phi_{31} & \phi_{32} & 1 & & \\ & .. & & .. & \\ & & .. & & .. \\ \phi_{p1} & \phi_{p2} & & \phi_{p,p-1} & 1 \end{vmatrix}$$

The p × p symmetric Φ has elements ϕ_{ij}, i,j = 1,2...p, which denote the covariance (correlations) between the common factors. Note that since each column of Λ may be scaled arbitrary, we have assumed without loss of

56

generality that the common factor has unit variance, and that is why the diagonal elements of Φ have been replaced with ones. If the assumption goes further to presume that factors are not correlated, then $\Phi = I$ and becomes $\Sigma_{xx} = \Lambda \Lambda' + \Psi$ .... (4-6).

Given particular Σ_{xx}, certain conditions must be met for the factorization in equation (4-6), and for it to be unique. The issue of identification is not easy to solve in this context. The total number of parameters required is the number of factor loadings, namely pq. There are $1/2p(p+1)$ separate variances and covariances in Σ_{xx}. Hence, we can determine by inspecting equation 4-6 that $1/2p(p+1)$, or $q<1/2p(p-1)$. Thus, q should be small compared to p. Equation 4-6 may be written using matrix notation as follows:

$$
\begin{vmatrix} X_1 \\ X_1 \\ .. \\ .. \\ X_p \end{vmatrix}
=
\begin{vmatrix} \lambda_{11} & .. & .. & .. & \lambda_{1q} \\ \lambda_{21} & \lambda_{22} & .. & .. & \lambda_{2q} \\ .. & .. & .. & .. & .. \\ .. & .. & .. & .. & .. \\ \lambda_{p1} & \lambda_{p2} & .. & .. & \lambda_{pq} \end{vmatrix}
\begin{vmatrix} f_1 \\ f_2 \\ \\ \\ f_p \end{vmatrix}
+
\begin{vmatrix} e_1 \\ e_1 \\ .. \\ .. \\ e_p \end{vmatrix}
$$

These equations are called factor patterns. The matrix that contains the correlations between the variables and the common factor is called the factor structure matrix or structure matrix. Both the pattern equation and the structural equation are necessary to get a complete solution.

Factors Solution

There are two solutions for the factor analysis model: the principal factor method, and the maximum likelihood method.

Principal Factor Method. Similarly to Principal Component Analysis, factors are extracted so that each factor accounts for the maximum possible amount of variance contained in the set of variables being factored. The difference is that in the Principal Factor Method each element of the correlation matrix is replaced by the respective variable's communality estimate (Dillon and Goldstein, 1984' Kim and Mueller, 1978).

Maximum Likelihood Procedure. The main objective of the maximum likelihood approach is to find the factor solution that best fits the observed correlations. When the study sample contains multivariate normal distributions of multiple variables, this procedure has the greatest likelihood of generating the observed correlation matrix. Because it is able to portray the maximum fit between the common factors and the observed variables, it is called the maximum likelihood method.

Application of Factor Analysis

Data for this example were derived from the National Health and Nutrition Examination Survey (HANES). A sample of 6,391 non-institutionalized adults received the detailed HANES examination, one component of which was the General Well Being (GWB) index. The GWB is a self-administered questionnaire that measures general health status. It contains questions about physical and psychological conditions.

Using principal component analysis with varimax rotation (Wan & Livieratos, 1978), the dimensions of the GWB index were uncovered for the study sample. The first common factor, accounting for 18.3 of the variance in GWB, is "depressive mood." Six GWB items having high lodgings on general well-being are: downhearted and blue; sad, discouraged or hopeless; nervousness; anxious, worried, or upset; under pressure; and afraid of losing mind or losing control. The second common factor, which is represented by seven GWB items, reflects health concerns. The third common factor is represented by five GWB items and reflects life satisfaction and emotional stability. Table 4 shows the three common factors and the items that load on each of them. This table shows that the three common factors account for 51.28 of the total variance in the general well-being (Wan & Livieratos, 1978). Next, an aggregate index of GWB was computed by summing the eighteen-item scores. The item-total score correlation coefficients were computed to determine the relevance of each GWB to the total configuration of general well-being.

Table 4. Three Factor Dimensions and Zero-Order Correlations between Each of the 18 General Well-being (GWB) Items and the Total GWB Scores

GWB Item	Factor loading	Correlation Coefficient
Depressive Mood (Eigenvalue): (3.300)		
Downhearted and blue	0.581	0.744
Sad, discouraged, hopeless	0.613	0.690
Anxious, worried, upset	0.733	0.717
Under stress, pressure	0.633	0.646
Nervousness	0.566	0.702
Afraid of losing mind or control	0.398	0.517
Health Concern (Eigenvalue):	(3.040)	
Bothered by bodily disorders	0.613	0.625
Health concern, worry	0.558	0.669
Feeling tired, worn-out	0.668	0.706
Waking up fresh, rested	0.537	0.639
Energy level	0.563	0.673
Good spirits	0.450	0.730
Relaxation	0.500	0.796
Life Satisfaction and Emotional		
Stability (Eigenvalue):	(2.89)	
Satisfied with life	0.529	0.587
Interesting daily life	0.629	0.590
Depressed, cheerful	0.536	0.751
Firm control of emotions	0.465	0.560
Emotionally stable	0.569	0.605
Total Percent of Variance:		
Explained	51.280	
GWB Score: Mean	80.340	
Standard deviation	17.676	

PATH ANALYSIS AND ITS APPLICATION

Path analysis is a methodology for analyzing systems of structural equations (Bollen, 1989) and empirically evaluating causal models. Path analysis was introduced by Sewall Wright (1934) and expanded by Hubert M. Blalock, Jr. (1971). Path analysis applies only to sets of relationships among variables that are linear, additive, and causal. The variables are assumed to be measurable on an interval scale.

Path analysis uses a structural equation model to specify the causal relationships among a set of variables, through path diagrams. A path diagram is a pictorial representation of a system of simultaneous equations. The main advantage of the path diagram is that it presents a picture of the relationships that are assumed to exist between the study variables. The actual construction of the causal model should be based on the researcher's knowledge of the subject matter and interpretation of current theory in his/her field. The researcher must state where relationships exist between two variables and the direction of each relationship. In essence, each specified relationship implicitly represents a hypothesis that can be tested by estimating the magnitude of the relationship.

To understand path diagrams, one must be familiar with the definitions of the symbols used. Most texts on path analysis use similar symbols. The following are some of the symbols that a researcher could use to construct a path diagram:

| X | An observed quantitative variable X

| Y | An observed quantitative variable Y

The observed variables or the variables that can be measured perfectly are enclosed in boxes. A straight single-headed arrow represents a causal relation between the variables connected by the arrow. Endogenous variables are those variables that are explained by the model, and are specified as causally dependent on other endogenous variables and/or exogenous variables. Exogenous variables are external to the system; they inform the variations in the endogenous variables.

Guidelines for Constructing a Path Diagram

The postulated causal relations among the variables of the system are represented by unidirectional arrows extending from each of the determining variables to each variable dependent on it.

X   Y

The postulated noncausal correlations <u>between exogenous</u> variables of the system are symbolized by a curved two-headed arrow.

X_1 X_2

Residual variables are represented by unidirectional arrows leading from the residual variable to the dependent variable.

In the path coefficient (P_{ij}), i denotes the dependent variable and j denotes the variable whose determining influence is under consideration. A path coefficient is also the standardized regression coefficient, which is derived by multiplying an unstandardized regression coefficient by a ratio of the standard deviation of an exogenous or predictor variable to the standard deviation of an endogenous variable. Residual path coefficient $= \sqrt{1 - R^2}$. The correlation of an exogenous variable and the dependent variable is the sum of the direct effect via the path coefficient from the exogenous variable to the dependent variable, and its indirect effect through its correlation with the other exogenous variable as measured by the product of the correlation and the path coefficient of the latter exogenous variable.

Path coefficients can be interpreted as the net change in the dependent variable affected by a one standard deviation change in a predetermined variable. Path analysis uses structural equations that represent the causal processes of the model to estimate the linkage between endogenous and exogenous variables, through the calculation of path coefficients. Path coefficients (b_{ij}) are standardized Ordinary Least Square (OLS) regression coefficients. The squared path coefficient (b^2) indicates the proportion of the variance of a dependent variable that the determining variable is directly responsible for (Land, 1969).

The path coefficients can be used to calculate a number of other statistics that are helpful in interpreting the causal system. The coefficient of determination (R^2) can be calculated for variables of interest, as the sum of such terms $P_{0i} r_{0i}$ where X_i is a determining variable. The *coefficient of alienation* (the residual) for the variable can then be calculated as $1-R^2$, and the square root of this is the residual path coefficient for the variable.

One of the main advantages of path analysis is that it can examine the direct and indirect effects of variables upon each other. A simple way to do this is to break down the correlation between two variables into the sum of simple and compound paths. The simple paths represent direct causal effects. Some of the compound paths represent indirect causal effects, and the others represent other indirect effects (Figure 11).

$$r_{01} = P_{01} + r_{12} P_{02}. \qquad \text{(Total effect)}$$
$$P_{01} \qquad \text{(Total direct effect)}$$
$$r_{12} P_{02} \qquad \text{(Total indirect effect)}$$

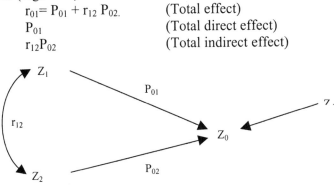

Figure 11. A Path Model

61

Z refers to the variable (standard form).
P refers to the path coefficient or standard partial regression coefficient.
Z_d refers to the residual variable.

Regression assumptions:

1. No specification error.
2. A linear relationship exists between X and Y.
3. No relevant independent variables have been excluded.
4. No irrelevant independent variables have been included.
5. No measurement errors: The variables X_i and Y_i are accurately measured.
6. Assumptions concerning the error term ε_i:
7. Zero mean: $E(\varepsilon_i) = 0$.
8. For each observation, the expected value of the error term is zero.
9. Homoscedastiscity: $E(\varepsilon_i^2) = $ Constant $=\sigma^2$, the variance of the error term is constant for all values of X_i.
10. No autocorrelation: $E(\varepsilon_i)(\varepsilon_j) = 0$ while $i \neq j$. The error terms are uncorrelated.
11. The independent variable is uncorrelated with the error term: $E(\varepsilon_i X_i)=0$.
12. Normality: the error term, ε_i, is normally distributed.

Procedures of structural equation:

1. Assuming correlation between two variables,
$$r_{ij}= \sum \frac{Z_i Z_j}{N} \text{ and } r_{jd} \neq 0.$$
2. Multiplying each endogenous variable by each preceding variable (j).
3. Assuming across all elements, and dividing by N to give the correlation.
4. Expanding equation and solving unknown elements in the equation.
5. Obtaining estimated correlation coefficients to compare with actual correlation coefficients for examining the overall validity of the model.
6. Calculating the indirect effect via paths.

Examples of Path Analytic Models

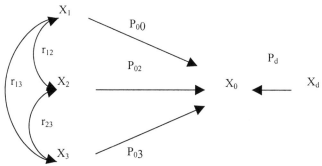

Figure 12. Path Analytic Model 1

Structural Equation:
$$X_0 = P_{01}X_1 + P_{02}X_2 + P_{03}X_3 + P_{0d}X_d \ldots\ldots\ldots\ldots\ldots(4\text{-}7)$$
If we consider X is measured in a standard form in Figure 12, so
$X_0 = Z_0;$
$X_1 = Z_1 ... X_3 = Z_3.$

$$r_{01} = \sum \frac{X_0 X_1}{N} \; ; r_{ij} = \sum \frac{Z_i Z_j}{N} \quad \text{and} \quad r_{jd} = 0.$$

Estimation Equations:

Using $\sum \dfrac{X_j}{N}$ times equation (4-7), we get the following:

$$r_{01} = P_{01}r_{11} + P_{02}r_{12} + P_{03}r_{13} + P_{0d}r_{1d}. \qquad (4\text{-}8)$$
$$r_{02} = P_{01}r_{12} + P_{02}r_{22} + P_{03}r_{23} + P_{0d}r_{2d}. \qquad (4\text{-}9)$$
$$r_{03} = P_{01}r_{13} + P_{02}r_{23} + P_{03}r_{33} + P_{0d}r_{3d}. \qquad (4\text{-}10)$$

Since r_{11}, r_{22}, and r_{33} equal 1, we drop these elements from the above equations. Then,

$$r_{01} = P_{01} + P_{02}r_{12} + P_{03}r_{13}. \qquad (4\text{-}11)$$
$$r_{02} = P_{01}r_{12} + P_{02} + P_{03}r_{23}. \qquad (4\text{-}12)$$
$$r_{03} = P_{01}r_{13} + P_{02}r_{23} + P_{03}. \qquad (4\text{-}13)$$

Solve equations 4-11 to 4-13 by the following:

$$\underset{\sim}{r_0} = \underset{\sim}{R}\underset{\sim}{P} = \underset{\sim}{R}\underset{\sim}{b}^*. \qquad (4\text{-}14)$$

P is a column vector of path coefficients. Note that P = b* (the vector of path coefficients is actually the vector of standardized partial regression coefficients); R refers to the observed correlation coefficients; and r_0 refers to estimated correlation coefficients. Equations could be written in the matrix format as follows:

63

$$
\begin{vmatrix} r_{01} \\ r_{02} \\ r_{03} \end{vmatrix}_{3*1} = \begin{vmatrix} r_{11}\ r_{12}\ r_{13} \\ r_{21}\ r_{22}\ r_{23} \\ r_{31}\ r_{32}\ r_{33} \end{vmatrix}_{3*3} \begin{vmatrix} P_{01} \\ P_{02} \\ P_{03} \end{vmatrix}_{3*1}
$$

To solve b* or path coefficient (P), we can derive from the above matrix, r_0: $\underset{\sim}{b} = \underset{\sim}{R}^{-1} \underset{\sim}{r_0}$, where R refers to correlation coefficients among the exogenous variables, and r_0 refers to correlations between the dependent (endogenous) variable and exogenous variables.

To use the following SAS program, one can invert a matrix to solve and estimate P:

```
PROC IML;
    PRINT 'R MATRIX';
    R= { 1.00 0.03 -0.01,
         0.03 1.00  0.60,
        -0.01 0.60  1.00};
X_NAMES={'X1' 'X2' 'X3' };
BLANK={'1', '2', '3'};
PRINT R [COLNAME=X_NAMES ROWNAME=BLANK];
r_0={0.06,
    -0.48
     0.06};
PRINT r_0 R [COLNAME=X_NAMES ROWNAME=BLANK];
PRINT 'INVERSE OF A MATRIX';
RINVVINV(R);
PRINT RINVV [COLNAME=X_NAMES ROWNAME=BLANK];
PRINT 'PATH COEFFICIENT';
RINVV=INV(R);
P=RINVV*r_0;
PRINT P [COLNAME=X_NAMES ROWNAME=BLANK];
```

The output of the program will be:
R MATRIX

R	X1	X2	X3
1	1	0.03	-0.01
2	0.03	1	0.06
3	-0.01	0.06	1

r_0 MATRIX

r_0	R	X1	X2
0.06	1	1	0.03
-0.48	2	0.03	1
0.06	3	-0.01	0.06

INVERSE OF A MATRIX

RINVV	X1	X2	X3
1	1.0021295	-0.005637	-0.04348432
2	-0.005637	1.5656708	-0.939966
3	-0.01	0.939966	1.5656708

Path Coefficient

P	X1
1	0.0898159
2	-0.811302
3	-0.5476794

Comparison between observed and estimated correlation coefficients:

Estimated r	Observed r
$R*_{01}$	r_{01}
$R*_{02}$	r_{02}
$R*_{03}$	r_{03}

For overidentified models, the sum of the paths can be used to validate the model by calculating each correlation and comparing it to the original correlation. If these are equal for all possible linkages, the model can be considered valid. If not, the model should be re-evaluated and perhaps revised. However, any revisions to the model should be based on the theory behind the model and not simply on the statistics generated by path analysis.

An Example of Path Analysis

$X_0 = b_1X_1 + b_2X_2 \ldots \ldots \ldots (4\text{-}15)$.

R= regular correlation
matrix of all three
variables included
(X_0, X_1, X_2)

	X_1	X_2	X_0
X_1	1.00	.316	.600
X_2	.316	1.00	.800
X_0	.600	.800	1.00

$= \ |\ R|\ \underline{r}_0\ |$

$|A| = 1 - r_{12} \bullet r_{21} = 1 - (.316)2 = .90$

$\underline{B}* \ R^{-1}\ r_0 = \begin{vmatrix} 1/.9 & -.316/.9 \\ \\ -.316/.9 & -1/.9 \end{vmatrix} \begin{vmatrix} .600 \\ \\ .800 \end{vmatrix} = \begin{vmatrix} .6/.9+ (.8(-.316))/.9 \\ \\ (.6(-.316))/.9-.8/.9 \end{vmatrix} = \begin{vmatrix} .386 \\ \\ .678 \end{vmatrix}$

So,

$$X_0 = .386X_1 + .678X_2 .$$

Then the submodel is viewed as

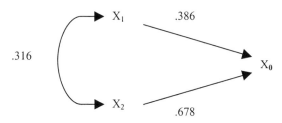

Additional Examples: Figure 13 presents a growth model with repeated measures of the same outcome variable. The equation is as follows:

$X_0 = b_1X_1 + b_2X_2 + b_3X_3 \ldots \ldots \ldots (4\text{-}15)$

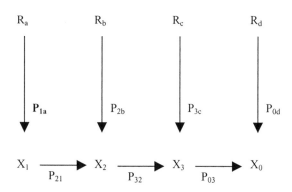

Figure 13. Path Analytic Model 3

Structural equations:

$$X_2 = P_{21} X_1 + P_{2b} R_b \qquad (4\text{-}15)$$
$$X_3 = P_{32} X_2 + P_{3c} R_c \qquad (4\text{-}16)$$
$$X_0 = P_{03} X_3 + P_{0d} R_d \qquad (4\text{-}17)$$

Estimation equations:

$$r_{12} = P_{21} = P_{21} r_{11} + P_{2b} r_{1b}. \qquad (4\text{-}18)$$

$$r_{13} = P_{32} r_{12} = P_{32} r_{12} + P_{3c} r_{1c}. \qquad (4\text{-}19)$$

$$r_{01} = P_{03} r_{13} = P_{0d} r_{1d} + P_{03} r_{13}. \qquad (4\text{-}20)$$

Multiplying $\Sigma\, X_2/N$ to equations (4-16) and (4-17), we get:

$$r_{23} = P_{32}. \qquad (4\text{-}21)$$

$$r_{02} = P_{03} r_{23}. \qquad (4\text{-}22)$$

Multiplying $\Sigma\, X_3/N$ to equation X_0, we get:

$$r_{03} = P_{03} r_{33} + P_{0d} r_{3d} = P_{03}. \qquad (4\text{-}23)$$

Comparison between estimated and observed correlation coefficients: we use equations (4-21), (4-22) and (4-23) to solve other three equations.

Estimated r	Observed r
$r^*_{03} = P_{32} r_{12} = r_{12}\, r_{12}$	r_{13}
$r^*_{01} = P_{03} r_{13} = r_{03}\, r_{13}$	r_{01}
$r^*_{02} = P_{03} r_{23} = r_{03}\, r_{23}$	r_{02}

If $r_{ij} = r^*_{ij}$, we find this causal model is valid. If there is a big discrepancy between the observed and estimated coefficients, we may find that there are correlations between the estimated variables (i.e. a, b, and c).

Figure 14 presents a path model that shows X_1 exerts both a direct effect and an indirect effect via X_2 and X_3.

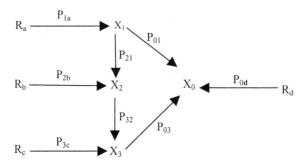

Figure 14. Path Analytic Model 3

Structural Equations:

$$X_2 = P_{21} X1 + P_{2b} R_b = P_{21} X_1. \qquad (4\text{-}24)$$
$$X_3 = P_{32}X_2 + P_{3c} R_c = P_{32} X_2. \qquad (4\text{-}25)$$
$$X_0 = P_{03}X_3 + P_{0d}R_d = P_{01}X_1 + P_{03}X_3. \qquad (4\text{-}26)$$

Estimation Equations:

$$r_{03} = P_{21}. \qquad\qquad r_{21} = P_{23}.$$
$$r_{13} = P_{32}\, r_{32}. \qquad\qquad r_{02} = P_{01}\, r_{12} + P_{03}\, r_{23}.$$
$$r_{01} = P_{01} + P_{03}\, r_{13}. \qquad r_{03} = P_{01}\, r_{13} + P_{03}.$$

Solve P_{ij} as follows: For example, to get P_{01} and P_{03}, we can use two equations:

$$r_{01} = P_{01} + P_{03}\, r_{13}.$$

$r_{03} = P_{01}\, r_{13} + P_{03}$, using r_{13} times equation r_{03} to get the following solution:

$$r_{13}\, r_{03} = P_{01} + P_{03}\, r_{13}.$$
$$r_{01} - r_{13}r_{03} = P_{01} - P_{01}\, (r_{13})^2.$$

We get $P_{01} = (r_{01} - r_{13}\, r_{03}) / (1 - r_{13}^2)$. Similarly, we apply the same procedure and get $P_{03} = (r_{03} - r_{13}\, r_{01}) / (1 - r_{13}^2)$.

CONCLUSION

The correlational data can potentially provide useful information to examine the structural relationships among the study variables. At the exploratory level, researchers can use statistical methods such as AID analysis, logistic regression analysis, factor analysis and path analysis to explore the data structure and generate the relational patterns of a complex set of indicators or observed variables. The structural relationship between exogenous and endogenous variables can be further postulated and specified in the model building or model generation stage. The plausibility can be further confirmed or disconfirmed by replicating the study or gathering new data from different samples to validate the proposed models.

In conclusion: preliminary evidence in health care management can be derived from the exploratory approach. The exploratory statistical methods can be effectively used to generate models guided by theoretical specifications. However, a theoretically based model must be validated by confirmatory statistical methods.

REFERENCES

Andersen, R., Newman, J. (1973). Societal and individual determinants of medical care utilization in the United States. *Milbank Memorial Fund Quarterly* 51: 95-124.

Andersen, R., Smedby, B., Anderson, O.W. (1970). *Medical Care Use in Sweden and the United States*. Chicago: University of Chicago, Center for Health Administration Studies.

Blalock, H.M. (1971). *Causal Models in the Social Science*. Chicago: Aldine-Atherton.

Bollen, K.A. (1994). *Structural Equations with Latent Variables*. New York, NY: John Wiley & Sons.

Dillon, W., Goldstein, M. (1984). *Multivariate Analysis: Methods and Applications*. New York: John Wiley and Sons.

Fetter, R.B., Freeman, J.L., Park, H., Schneider, K., Lichtenstein, J. and Health System Management Group. (1989). *DRG Refinement with Diagnostic Specific Morbidities and Complications: A Synthesis of Current Approaches to Patient Classification* (Final Report). New Haven, CT: School of Management, Yale University.

Fries, B. E., Schneider, D. P., Foley, W. J., Gavazzi, M., Burke, R., Cornelius, E. (1994). Refining a case-mix measure for nursing homes: Resource utilization groups (RUG-III). *Medical Care* 32: 668-685.

Hartwig, F., Dearing, B. E. (1979). *Exploratory Data Analysis*. Beverly Hills, CA: Sage Publications.

Health Care Financing Administration. (2000). HCFA.GOV/MEDICARE/IRFPPS.HTML.

Hosmer, D.W., Lemeshow, S. (1989). *Applied Logistic Regression*. New York: John Wiley & Sons.

Kim, J.O., Mueller, C.W. (1978). *Factor Analysis: Statistical methods and practical issues*. Beverly Hills, CA: Sage Publications.

Land, K. (1969). Principles of path analysis. In E.F. Borgatta (ed.), *Sociological Methodology*. San Francisco: Jossey-Bass, pp. 3-37.

Sonquist, J.A., Baker, E.L., Morgan, J.N. (1973). *Searching for Structure*. Ann Arbor, Michigan: University of Michigan Institute for Social Research.

Steinberg, D., Colla, P. (1995). *CART: Three-Structured Nonparametric Data Analysis*. San Diego, CA: Salford Systems.

Wan, T.T.H., Soifer, S. (1974). Determinants of physician utilization: A causal analysis. *Journal of Health and Social Behavior* 15: 100-108.

Wan, T.T.H., Yates, A. Prediction of dental services utilization: A multivariate approach. *Inquiry* 12: 143-156.

Wan, T.T.H., Livieratos, B.B. (1978). Interpreting a general index of subjective well-being. *Milbank Memorial Fund Quarterly* 56: 531-556.

Wan, T.T.H., Weissert, W.G., Livieratos, B.B. (1980). Geriatric day care and homemaker services: An experimental study. *Journal of Gerontology* 35: 256-274.

Wan, T.T.H. (1989). The effect of managed care on health services use by the dually eligible elders. *Medical Care* 27: 983-1001.

Wan, T.T.H. (1995). *Analysis and Evaluation of Health Care Systems: An Integrated Approach to Managerial Decision Making*. Baltimore, MD: Health Professions Press.

Wright, S. (1934). The method of path coefficients. *Annals of Mathematical Statistics* 5: 161-215.

CHAPTER 5

INTRODUCTION TO STRUCTURAL EQUATION MODELING

In this chapter, we introduce structural equation modeling, or the analysis of Linear Structural Relations (LISREL).

THE STRUCTURAL EQUATION MODEL

The structural equation model is an extension of regression methods. Karl Jöreskog and colleagues developed a generic statistical modeling of multivariates, using analysis of covariance structural models (Jöreskog, K. G. and D. Sörbom , 1979). This work has inspired many researchers in the social sciences to take advantage of the structural equation models with unobservable (latent) constructs for parameter estimation and hypotheses testing in causal models (Dillon & Goldstein, 1984). The structural equation model is like the path-analytic model in three ways: (1) model construction, (2) parameter estimation of the model, and (3) testing the fit of the model to the data by comparing observed correlations with predicted correlations among the study variables. However, the LISREL model is less restrictive than the path-analytic model. For instance, it allows the investigator to ask direct questions about the data, in the form of different restrictions on the coefficients. LISREL can easily handle errors in measurement, correlated errors and residuals, and reciprocal causation. Thus, LISREL's advanced procedures can specify, compare, and evaluate the impact of an intervention on a set of correlated outcome variables (Bollen, 1989; Jöreskog, K. G. and D. Sörbom, 1979).

The LISREL model actually contains two parts. One is the measurement model, and the other is the structural equation model. In the following sections of this book, a chapter is allocated to each model, to provide a full, formal presentation of the models. Chapter 6 discusses the measurement model in the context of confirmatory factor analysis, and Chapter 7 presents the structural equation model.

Definition of Structural Equation Modeling

The following path diagrams, in conjunction with a few rules, allow us to express the values of a dependent variable in terms of values of its source variables.

Rule One. The value of a variable determined by only a single source is the value of the source times the structural coefficient (A).

$$A$$

X $\longrightarrow$ Y, implies that Y = A * X.

$$5 \quad \swarrow$$

X $\longrightarrow$ Y, implies that Y = A * X + 5 (residual term).

Rule Two. The value of a variable determined by two or more sources is the sum of source values, each multiplied by its respective structural coefficient. The order of the summation does not matter.

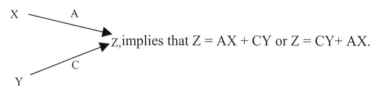

Z, implies that Z = AX + CY or Z = CY + AX.

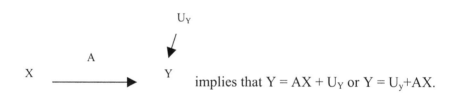

implies that Y = AX + U_Y or Y = U_y + AX.

Structural Equation Presentations

The above rules define how verbal statements of theory are converted to a flow graph formulation, and how the flow graph defines mathematical (functional) relations among the study variables. The structural relations can be graphically presented to reveal the causal processes under study (Bentler, 1988). Structural-equation formulations are important in more advanced treatments of complex systems.

LISREL is a statistical modeling tool for the analysis of data according to specified causal models and systems of structural equations. LISREL, which stands for Linear Structural Relations, is based on a general model that comprises: A measurement model specifying the relations between the observed variables and unobserved, theoretical variables or latent variables including measurement errors. A linear structural equation model specifying causal relationships among the study variables with possibly reciprocal causation and correlated random disturbance terms.

The LISREL model is very flexible. It includes as sub-models many standard and non-standard models such as measurement models, confirmatory factor analysis models, and recursive and non-recursive path analytic models.

A specified model may be estimated and tested by the LISREL (Jöreskog and Sörbom, 1993), the AMOS (Arbuckle and Wothke, 1999), the Mplus (Muthén and Muthén, 1998), the EQS (Bentler and Wu, 1994), the STREAMS (Gustafsson and Stahl, 1999), or the Mx (Neale, et al., 1999) computer program, which will fit the moment structure implied by the model to the moment matrix of the observed variables. Structures on the means of the observed variables can also be specified and fitted.

The parameters of the model may be estimated by different methods including the maximum likelihood method, and measures of goodness of fit are provided by the computer program. The program can analyze data from several independent samples according to models in which some parameters are specified to be invariant across populations. Analysis of ordinal data can be accomplished by the computation and analysis of polychoric (binary categorical) and polyserial (ordered-categorical) correlations that are measures of bivariate association, arising when one or both observed variables are categorical variables (Drasgow, 1988).

Two Main Types of Models

The following questions can explore the relationships among study variables: Can X_1, X_2 and X_3 adequately measure the theoretical variable ξ (Figure 15)? If so, how reliable is each of the X_i? What is the (causal) relationship between the theoretical variables (Figure 16)? How strong is this relationship?

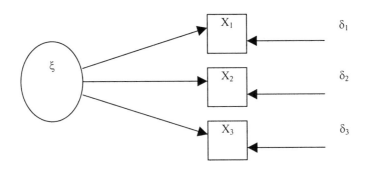

Figure 15. A Measurement Model

Figure 16. A Structural Model

Examples of Theoretical Construct Variables and Congeneric Measures

Researchers in health services management have used health status and functional measures as outcome variables. Multiple indicators for each functional status are used in formulating measurement models (Figure 17 and Figure 18). The effect of service use or medical intervention can then be examined (Figure 19).

Construct   **Congeneric Measures**

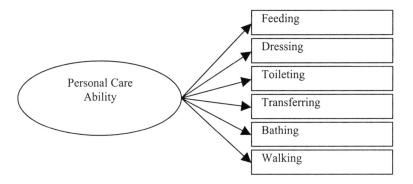

Figure 17. Model 1: Physical Functioning

Construct   **Congeneric**

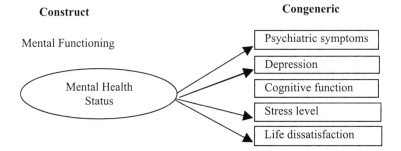

Figure 18. Model 2: Mental Functioning

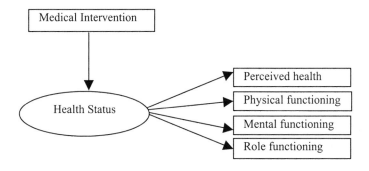

Figure 19. Health Services Research Models

75

It is important to note that a latent construct is conceptualized as a theoretical variable that may be measured by multiple indicators. For example, a measurement model of patient care outcomes at the organizational (hospital) level can be developed with multiple indicators of adverse health events: trauma rate (TRUMAR), rate of discharges with unstable medical conditions (MEDPROBR), rate of treatment problems (TXPROBR), postoperative complication rate (COMPRATE), and rate of unexpected deaths (DEDPROB) (Figure 20). This model assumes that adverse health events at hospitals are positively correlated. Concomitant occurrences of poor outcomes indicate poor quality of care provided by a hospital.

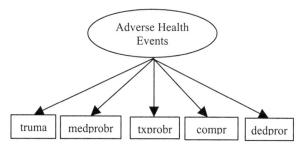

Figure 20. A Measurement Model of Patient Adverse Outcomes

Basic methodological considerations in structural equation modeling are the following:

1. Linearity: Linear relationships exist among the variables studied.
2. Additivity: The linear effects of multiple predictor variables on a criterion variable are additive.
3. Causal priority: The causal sequelae are logically determined, so that an end result should occur after the presence of a causative factor.
4. Interval, ordinal, and nominal scales: Measurement scales vary, but continuous variables are desirable in structural equation modeling.
5. Independence and identical distributions of observations: Observations are assumed to be independent.
6. Homoscendasticity: There is constant or equal variance in an outcome variable across all values of the explanatory variables in linear regression or analysis of variance.
7. Uncorrelated vs. correlated residual terms or disturbances: Residuals of different equations may be assumed to be uncorrelated or correlated.
8. Multicollinearity: Highly correlated variables used in the measurement model may cause multicollinearity problems that will generate biased estimates of the parameters.

9. Standardization: Parameters can be standardized so that relative weights of the indicator variables can be determined.
10. Reliability and validity: Both measurement issues can be analyzed in structural equation modeling.
11. Goodness of fit: Statistical measures are generated to evaluate the overall model fit as well as the statistical significance of the variables.
12. Predictive validity: The predictive power of a measurement instrument for health status can be evaluated in a panel study design to determine the stability of the measurement model for health status over time, as well as its predictability for a future occurrence (e.g., consumption of health services).

The LISREL model assumes that one specifies a causal structure among a set of latent variables or hypothetical constructs, some of which are designed as dependent and others as independent variables (Jöreskog, 1978). These latent variables may not be directly observed, but a set of observed variables is related to the latent variables. Thus, the latent variables appear as underlying causes of the observed variables. To familiarize the reader with the mathematical symbols used in LISREL, Table 5 presents the Greek alphabet.

Table 5. The Greek Alphabet

	Form		Name	Sound		Form		Name	Sound
	Capital	Small				Capital	Small		
1	A	α	alpha	a	13	N	ν	nu	n
2	B	β	beta	b	14	Ξ	ξ	xi	x
3	Γ	γ	gamma	g	15	O	o	omicron	o
4	Δ	δ	delta	d	16	Π	π	pi	p
5	E	ε	epsilon	e	17	P	ρ	roo	r(rho)
6	Z	ζ	zeta	z	18	Σ	σ	sigma	s
7	H	η	eta		19	T	τ	tau	t
8	Θ	θ	theta	th	20	Y	υ	upsilon	u
9	I	ι	iota	i	21	Φ	φ	phi	ph
10	K	κ	kappa	k	22	X	χ	khi	kh
11	Λ	λ	lambda	l	23	Ψ	ψ	psi	ps
12	M	μ	mu	m	24	Ω	ω	omega	o

Measurement Models

Most theories and models in the social and behavioral sciences are formulated in terms of hypothetical concepts and constraints, so-called latent variables, which are not directly measurable (observable), but are often measured more or less well by a number of indicators (e.g., symptoms). The purpose of the measurement model is to describe how well the observed indicators work as a measurement instrument for the latent variables.

Measurement Errors

Because most measurements used in the investigation contain sizable measurement errors, unless they are taken into account there may be biased results. Measurement errors occur because of imperfections in measurement instruments (questionnaires, interviews, tests, etc.) and procedures (recording, coding, scaling, grouping, aggregation, etc.). Often, although a number of different and potentially equivalent measures could be used to measure the same thing, the actual observed values for the variables may not be the same. Thus, adequate modeling should take measurement error into account whenever possible. In illustrating the measurement model, the following conventions for path diagrams are used: A one-way arrow between two variables indicates a postulated direct influence of one variable on another. A two-way arrow between two variables indicates that these variables are correlated. Coefficients are associated to each arrow as follows. Arrows from (ξ)-variables to x-variables are denoted (λ), arrows from one (ξ) to another (ξ) are denoted (ϕ). Each coefficient has two subscripts, the first being the subscript of the variable that the arrow is pointing to, and the second being the subscript of the variable that the arrow is coming from. For two-way arrows the two subscripts may be interchanged, so that (ϕ_{21}) = (ϕ_{12}) in Figure 21. Arrows that have no coefficients in the path diagram are assumed to have a coefficient of one.

All direct influences of one variable on another must be included in the path diagram. Hence the absence of an arrow between two variables means that it is assumed that these two variables are not directly related.

If the above conventions for path diagrams are followed, it is always possible to write the corresponding equations by means of the following rules:

1. For each variable that has a one-way arrow pointing to it, there will be one equation in which this variable is a left-hand variable (e.g., $X_1 = \lambda_{11} \xi_1 + \delta_1$).

2. The right-hand side of each equation is the sum of a number of terms equal to the number of one-way arrows pointing to that variable, and each term is the product of the coefficient associated with the arrow and the variable from which the arrow is coming.

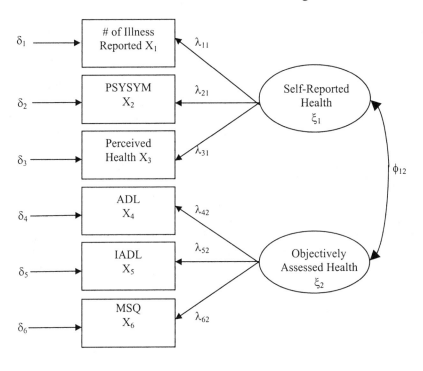

Figure 21. A Measurement Model of Health Status

CONFIRMATORY FACTOR ANALYSIS

Confirmatory Factor Analysis (CFA) attempts to explain the variation and covariation in a set of observed variables in terms of a set of unobserved factors. The observed variables are conceptualized as linear functions of one or more factors. These factors can be either common (latent), meaning that they may directly affect more than one of the observed variables, or unique (residual), meaning that they may directly affect only one observed variable (Long, 1983). The relationship between the observed variables and these two types of factors can be expressed mathematically as:

$$X = \Lambda \xi + \delta_X \dots\dots\dots\dots\dots (5.1),$$

Where:

X is a (q * 1) vector of the observed variables;

ξ (xi) is a (q * 1) vector of the common factors;

Λ (lambda) is a (q * s) matrix of factor loading relating the observed variables X's to the latent variables ξ's; and

δ (delta) is a vector of the unique factors (measurement error δ terms).

The CFA is based on three underlying assumptions. First, it is assumed that both the latent and the observed variables are measured as deviations from their means. Second, it is assumed that the number of observed variables in X is greater than the number of latent factors in ξ. Third, it is assumed that the common factors and the unique factors (measurement errors) are uncorrelated.

The main advantage of confirmatory over exploratory factor analysis is that a researcher can impose substantively motivated constraints on the model. The constraints determine which pairs of common factors are correlated, which observed variables are affected by a unique factor, and which pairs of unique factors are correlated. Another advantage of CFA is its hypothesis - testing capability. A chi-square likelihood ratio test is calculated for the null hypothesis that the sample covariance matrix S is drawn from a population characterized by the hypothesized covariance matrix Σ. If the null hypothesis is not rejected, the sample data are considered to be consistent with the constraints imposed by the researcher, and the substantively generated model is confirmed.

STRUCTURAL EQUATION MODEL

The structural equation modeling technique was elaborated by Karl Jöreskog and his colleagues. They developed a general statistical model and a computer program for the analysis of covariance structural models (Jöreskog and Söborm, 1979; Byrne, 2001).

Putting things together in the structural model, the covariance among the observed variables is decomposed in two conceptually distinct steps. First, the observed variables are linked to unobserved or latent variables through a pair of factor analytic models. Second, the causal relationships among these latent variables are specified through a structural equation model. The covariance structural model therefore consists of the simultaneous specification of factor models and a structural equation model.

The factor models that express the mathematical relationship between the observed variables and the latent variables, for the latent exogenous (ξ) variables and endogenous (η) variables are:

$$X = \Lambda_x \xi + \delta \text{ and } Y = \Lambda_y \eta + \varepsilon,$$

where:

x is a vector of observed exogenous variables;

y is a vector of observed endogenous variables;

Λ_x is a matrix of the loadings of the observed x variables on the latent ξ variables; and, similarly,

Λy is a matrix of the loadings of the observed Y variables on the latent η variables, δ and ε are vectors of unique factors.

Unique factors may be correlated only within each factor model, not across models.

The structural components of the covariance model are expressed mathematically as:

$$\eta = \beta \eta + \Gamma \xi + \zeta \dots \dots \dots \dots (5.2),$$

where:

η is a vector of latent endogenous variables;

ξ is a vector of latent exogenous variables;

B is a matrix of coefficients relating the endogenous variables one to another;

Γ is a matrix of coefficients relating the exogenous variables to the endogenous variables; and

ζ is a vector of errors in equations, indicating that the endogenous variables are not perfectly predicted by the structural equations.

X and Y are observed variables, ξ is an exogenous underlying construct or independent latent variable, and η is an endogenous underlying constraint or dependent latent variable (Figure 22).

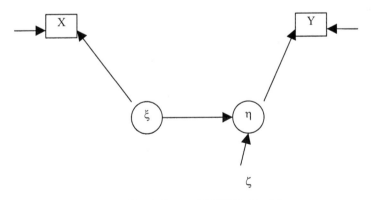

Figure 22. A General LISREL Model

The covariance structure model is based on five assumptions. First, it is assumed that the variables are measured from their means. Second, common and unique factors are assumed to be not correlated. Third, it is assumed that unique factors and residuals in equations are uncorrelated across equations. Fourth, exogenous variables and residuals in equations are assumed to be uncorrelated. Lastly, it is assumed that none of the structural equations is redundant or duplicative.

LISREL has hypothesis - testing and model fitting capabilities. It can be applied to validate both the measurement models and the entirely specified model with structural equations. An X^2 likelihood ratio test is calculated for the null hypothesis that the sample covariance matrix is drawn from a population characterized by the hypothesized covariance matrix (Σ). If the null hypothesis is not rejected, the adequacy of the model specified by the researcher is confirmed.

After model specification is performed, the assessment of model fit is undertaken to ensure the appropriate interpretation of the theoretical framework (Bollen, 1989; Maruyama, 1998). The criteria for assessment of fit are examination of the solution, measures of overall fit, and detailed assessment of fit. In the first step, to check for the appropriateness of each variable. Parameter estimates with the right sign and size, standard errors within reasonable ranges, correlations of parameter estimates, and squared multiple correlations are commonly used. In the second step, the overall model fit is evaluated to see how well the specified model fits the data. The indicators are goodness-of-fit index (GFI), adjusted goodness-of-fit (AGFI), root mean squared error of approximation (RMSEA), related significance statistics (P-close), and Hoelter's critical N (C.N.). The third step is to identify the possible sources of the lack of fit. The commonly used indicators are modification indices, which show the extent to which the model fit could be improved by adding certain constraints between variables. Briefly, GFI (goodness-of-fit, ranging from 0 to 1) is a measure of the amount of variances and covariances jointly accounted for by the model; the larger, the better. AGFI (adjusted goodness-of-fit) is a measure of goodness-of-fit while taking into account the degrees of freedom available. RMSEA (root mean squared error of approximation) measures the degree of model adequacy as based on population discrepancy in relation to degrees of freedom; a value less than 0.05 (or 0.08) is acceptable. P-close is a "p-value" for testing the null hypothesis that RMSEA is equal to or less than 0.05. A P-close value equal to or greater than 0.05 indicates a close model fit. CN (Hoelter's critical N) indicates the largest sample size for which one would accept the hypothesis that a model is correct. Usually, a CN equal to or greater than 200 is needed.

An Application

The following example illustrates how a measurement model of adverse patient care outcomes (Figure 23) is developed and validated. "Adverse health outcomes," a latent variable, regresses on a set of indicator variables to show how hospital performance (i.e. adverse patient outcomes) is influenced by a variety of exogenous variables. "Adverse patient outcomes" (patients with specific unfavorable results of hospitalization) is an unobserved, or latent, construct. To develop a measurement model, Wan (1992) selected five indicators of adverse patient outcomes, using the hospital as the unit of analysis. The indicators are in-hospital trauma rate (TRAUMAR), rate of discharges with unstable medical conditions (MEDPROBR), rate of treatment problems (TXPROBR), postoperative complication rate (COMPRATE), and rate of unexpected deaths (DEDPROB). The measurement model (Figure 23) specifies the relationship between the five observed indicators and the unobserved theoretical construct (i.e., adverse patient outcomes).

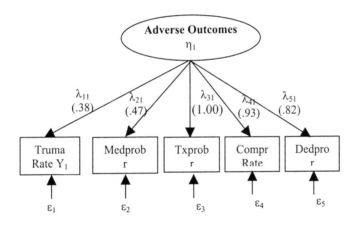

Figure 23. A Measurement Model of Adverse Health Care Outcomes in Acute Care Hospitals

Each indicator is considered a linear function of the common factor, adverse patient outcomes. The common factor (η) may directly affect more than one observed indicator. A measurement error term (unique factor or ε) is associated with each specific outcome indicator (y). The common factor is not correlated with the unique factors. However, the measurement errors (ε_s) may be correlated with each other. Therefore, the relationship between the observed indicators and the common factor can be expressed as $y = \lambda_y \eta + \varepsilon$; where y is a vector of observed outcome variables. Lambda y (λ_y) is a matrix of factor loadings relating the observed y's to the common factor (eta), and epsilons (ε) are the residual or unique factors. A change in eta (η) will result in a direct change in y. It is also assumed that the number of observed variables in y is greater than the number of common factors.

After a model has been specified, the next step is identification of the model, which means deciding whether there is a unique solution for each of the parameters of the model. If the number of unknowns is greater than the number of known parameters, the parameters cannot be estimated. In that case, respecification of the measurement model is necessary.

Guided by the original theoretical model, three steps can improve the goodness of model fit. The first step is to eliminate observed variables (indicators) that do not make a statistically significant contribution to the latent variable. The second step is to add other related indicators that are appropriate measures of the latent variable. The third step is to free the parameter (lambda or epsilon) with the largest modification index generated by the LISREL program within constraints imposed by the theoretical model.

THE CONVARIANCE STRUCTURE MODEL

By using validated measures of a latent variable such as adverse patient outcomes, we can further examine the effect of such interventions as total quality improvement programs and incentive programs on adverse outcomes. The specification of this structural equation model is: adverse outcomes = f (interventions) + residual term. If researchers deal with confounding or extraneous variables in the causal analysis, those variables must be included as control variables in the structural equation in order to determine the net effect of an intervention on outcomes. This model can be validated through covariance structure analysis, which provides parameter estimates simultaneously for the measurement model and the structural equation model. In our example, the explanatory variables were hospital bed size, number of high technology services offered, case mix, severity of patients treated, ownership, cost, technical efficiency, average length of stay, market share, net profit, and metropolitan size. Adverse outcomes, the latent construct, was regressed on thirteen predictor variables. Parameter estimates were derived using maximum likelihood methods for both measurement and structure equation models.

The results for this example are shown in Table 6. The explanatory variables account for 39.8 percent of the total variance in hospitals' adverse patient outcomes. The average length of stay is positively related to adverse outcomes when other explanatory variables are simultaneously controlled. Technical efficiency is negatively associated with adverse outcomes; the higher the degree of technical efficiency, the lower the level of adverse outcomes. The analysis shows that hospitals' problems with quality of care can be explained by hospital operational factors.

CONCLUSION

Structural equation modeling is a power analytical tool to validate the plausibility of a theoretically assumed structure of a set of the study variables, including exogenous and endogenous variables. Structural equation modeling can offer a distinct opportunity to examine factors affecting health services use and/or health care outcomes. Furthermore, this analytical strategy can help capture the causal processes of the variables under study.

Table 6. The Net Effects of Predictors on Adverse Patient Outcomes

Predictor Variables													
Dependent variable	BED SIZE	HITECH	CASE MIX	SEVER-ITY	METRO SIZE	MULTI-SYS-TEM	MD SCHL	FOR-PROFIT	COST	ALOS	EFFIC-IENCY	% SHARE	NET PROFIT
Outcomes	-0.170	-0.038	-0.070	0.071	0.083	-0.002	-0.110	-0.072	-0.128	0.230	-0.447	-0.120	-0.043
T-value	-0.767	-0.222	-0.437	0.758	0.631	-0.015	-0.835	-0.687	-1.103	1.986*	-4.611*	-1.061	-0.412

* Significant at 0.05 or lower level

$R^2 = 0.398$.

Goodness of fit (GOF) Statistics: $\chi2 = 78.330$ with 57 degrees of freedom; P = 0.049. GOF index = 0.918.

Adjusted GOF index = 0.755.

Root mean squares residual = 0.058.

REFERENCES

Arbuckle, J.L., Wothke, W. (1999). *AMOS 4.0 User's Guide.* Chicago: SamllWaters Corporation.

Bentler, P.M. (1988). Causal modeling via structural equation systems. In J.R. Nesselroade and R.B. Cattell (eds.). *Handbook of Multivariate Experimental Psychology* (2nd edition). New York: Plenum.

Bentler, P.M., Wu, E.J.C. (1993). *EQS/Windows User's Guide: Version 4.* Los Angeles: BMDP Statistical Software.

Bollen, K.A. (1989). *Structural Equations with Latent Variables.* New York, NY: John Wiley & Sons.

Byrne, B.M. (2001). *Structural Equation Modeling with AMOS.* Mahwah, NJ: Lawrence Erlbaum Associates, Inc.

Dillon, W., Goldstein, M. (1984). *Multivariate Analysis: Methods and Applications.* New York: John Wiley and Sons.

Drasgow, F. (1988). Polychoric and polyserial correlations. In Kotz, L., Johnson, N.L (Eds.), *Encyclopedia of Statistical Sciences* 7: 69-74. New York: Wiley.

Gustafsson, J., Stahl, P.A. (2000). *STREAMS User's Guide. Version 2.1. for Window.* Angered, Sweden: Multivariate Ware.

Jöreskog, K.G. (1978). Structural analysis of covariance and correlation matrices. *Psychometrika* 43: 443-477.

Jöreskog, K.G., Sörbom, D. (1979). *Advances in Factor Analysis and Structural Equation Models.* Cambridge, MA: Abt.

Jöreskog, K.G., Sörbom, D. (1993). *LISREL 8: Structural Equation Modeling with the SIMPLIS Command Language.* Chicago: Scientific Software International, Inc.

Long, J. S. (1983). *Covariance Structure Models.* Beverly Hills, CA: Sage Publications.

Long, J. S. (1983). *Confirmatory Factor Analysis.* Beverly Hills, CA: Sage Publications.

Maruyama, G. M. (1998). *Basics of Structural Equation Modeling.* Thousand Oaks, CA: Sage Publications.

Muthén, L. K., Muthén, B.O. (1998). *Mplus: The Comprehensive Modeling Program for Applied Researchers.* Los Angeles, CA: Muthén & Muthén.

Neale, M.C., Boker, S.M., Xie, G., Maes, H.H. (1999). *Mx: Statistical Modeling.* Richmond, VA: Virginia Institute for Psychiatric and Behavioral Genetics, Virginia Commonwealth University.

Wan, T.T.H. (1992). Hospital variations in adverse patient outcomes. *Quality and Utilization Review* 7(2): 50-53.

CHAPTER 6

CONFIRMATORY FACTOR ANALYSIS

In this chapter, we present LISREL's measurement model in the context of confirmatory factor analysis (CFA).

THE CFA CONTEXT

CFA attempts to explain the variation and covariation in a set of observed variables in terms of a set of theoretical, unobserved factors. The observed variables are conceptualized as linear functions of one or more factors. These factors can be either common (latent) factors which may directly affect more than one of the observed variables, or unique (measurement error) factors, which may directly affect only one observed variable (Long, 1983). The relationship between the observed variables and these two types of factors can be expressed mathematically as

$$X = \Lambda_x \xi + \delta \quad\dots\dots\dots\dots\dots\dots\dots (6.1),$$

where:

X is a (q * 1) vector of the observed variables;

Λ (lambda) is a (q * s) matrix of factor loadings relating the observed variables X's to the latent variables ξ's; and

δ (delta) is a vector of the unique factors.

The CFA is based on three underlying assumptions (Bollen, 1989; Long, 1983; Jöreskog and Sörbom, 1979). First, it is assumed that both the latent and the observed variables are measured as deviations from their means. Second, it is assumed that the number of observed variables in X is greater than the number of latent factors in ξ. Third, it assumed that the common factors and the unique factors are uncorrelated. The main advantage of confirmatory over exploratory factor analysis is that a researcher can impose substantively meaningful constraints on the model. The constraints determine which pairs of common factors are correlated, which observed variables are affected by a unique factor, and which pairs of unique factors are correlated.

Another advantage of CFA is its hypothesis-testing capability. A chi-square likelihood ratio test is calculated for the null hypothesis that the sample covariance matrix S is drawn from a population characterized by the hypothesized covariance matrix Σ. If the null hypothesis is not rejected, the sample data are considered to be consistent with the constraints imposed by the researcher, and the substantively generated model is confirmed.

Basic Confirmatory Factor Analysis Model

$$X = \Lambda x \zeta + \delta \qquad \text{Factor equation}$$

$$\Sigma = \Lambda_x \Phi \Lambda_x' + \Theta_\delta \qquad \text{Covariance equation}$$

$\Lambda_x \; \Phi \; \Theta_\delta$ may contain fixed, free, or constrained elements.

Θ_δ need not be diagonal (error terms may be correlated).

Assumptions
A. It is assumed that both the latent and the observed variables are measured as deviations from their means.
B. It is assumed that the number of observed variables in X is greater than the number of latent factors in ξ.
C. It assumed that the common factors and the unique factors are uncorrelated.

EXAMPLES OF MEASUREMENT MODELS

Types of Measurement Models

Figure 24 shows a latent construct, health care outcomes, that is reflected by four indicators of physical, mental and social functioning: 1) perceived health, 2) ADL functioning, 3) mental functioning, and 4) social functioning. This is a congeneric model because no constraints are imposed on factor loadings of the indicators.

Several examples are presented to illustrate the measurement models.

90

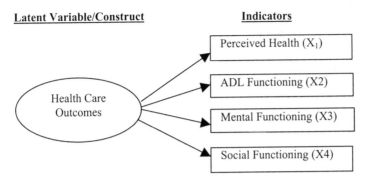

Latent Variable/Construct   Indicators

Perceived Health (X₁)

ADL Functioning (X2)

Health Care
Outcomes

Mental Functioning (X3)

Social Functioning (X4)

Figure 24. Measurement Model of Health Status/Outcomes

A Congeneric Measurement Model with Multiple Indicators

Figure 24 shows a common factor (ξ_1) shared by four related indicators with their respective measurement errors (deltas or δ's) with no constraints imposed on factor loadings. For example, competitiveness (ξ_1) of a health care market for acute care services is measured by four related indicators: the number of short-term general hospitals (X_1), the number of managed care organizations (X_2), the percentage of population enrolled in prepaid group practices or HMOs (X_3), and the number of patients on the waiting list for elective surgeries (X_4) in a service market area. The linkages between the indicators and their common factor or latent construct are lambda coefficients.

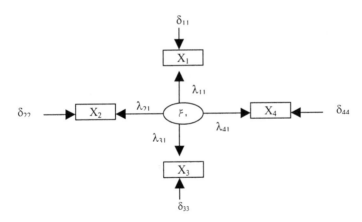

Figure 25. A Congeneric Measurement Model

91

$$
\begin{vmatrix} X_1 \\ X_2 \\ X_3 \\ X_4 \end{vmatrix} = \begin{vmatrix} \lambda_{11} \\ \lambda_{21} \\ \lambda_{31} \\ \lambda_{41} \end{vmatrix} \xi + \begin{vmatrix} \delta_{11} \\ \delta_{22} \\ \delta_{33} \\ \delta_{44} \end{vmatrix} = \underset{\sim}{\Lambda}\underset{\sim}{\xi} + \underset{\sim}{\delta}.
$$

4×1   4×1   4x1

Identification condition: Var $(\xi) = 1$.

$$
\Sigma = \begin{vmatrix} \lambda_1^2 + \theta_1 & & & \\ \lambda_2\lambda_1 & \lambda_2^2 + \theta_1 & & \\ \lambda_3\lambda_1 & \lambda_3\lambda_2 & \lambda_2^2 + \theta_1 & \\ \lambda_4\lambda_1 & \lambda_4\lambda_2 & \lambda_4\lambda_3 & \lambda_4^2 + \theta_4 \end{vmatrix} = \underset{\sim}{\Lambda}\underset{\sim}{\Lambda}' + \underset{\sim}{\Theta\delta}.
$$

Identification of this congeneric measurement model can be made as follows:

Degrees of freedom = number of known elements - number of parameters estimated = $[4\mathrm{x}(4+1)]/2 - 8 = 10 - 8 = 2$.

A Measurement Model with Two Correlated Theoretical Constructs

Figure 26 illustrates two correlated theoretical constructs in organizational research: 1) Organizational complexity (ξ_1) is measured by size (X_1) and number of divisions or subunits (X_2) within an organization. 2) Organizational specialization (ξ_2) is measured by the number of specialties (X_3) of professional staff and number of high-tech services offered. Each indicator has its measurement error (δ), indicating the imperfection of the measurement indicator. The measurement errors are not to be correlated. The relationships between the indicators and their theoretical construct are lambda coefficients (λ's). The two organizational constructs are assumed to be correlated (ϕ_{21}).

92

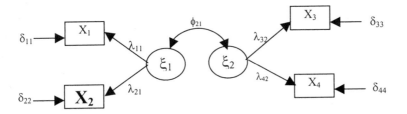

Figure 26. Two Sets of Congeneric Models

A Measurement Model with Three Correlated Constructs

The same principle can be applied to formulate a measurement model with three correlated theoretical constructs. Figure 27 shows the three mechanisms for establishing health informatic integration, which refers to the attainment of integration of administrative (ξ_1), managerial (ξ_2), and clinical (ξ_3) decision support systems within a health care organization. Each of the decision support systems has two or more indicators to measure the functional integration in its operations.

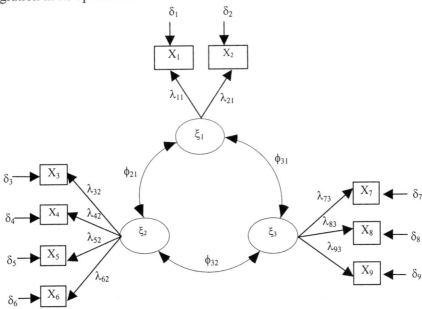

Figure 27. Three Sets of Congeneric Measures

In formulating a sound measurement model for health services management research, it is important to base the model on a theoretical specification. Several pitfalls should be avoided. One is classified the constructs without paying attention to their substantive meaning. For example, the concept of "organizational characteristics" is a broad classification of many phenomena such as the location of the organization in a rural or urban setting, ownership, size, age, and so on. Because many constructs are imbedded within the category of organizational characteristics, it is inappropriate to conceive that "organizational characteristics" is a latent construct. To avoid labeling a theoretical construct inaccurately, it is important to generate the measurement model under a specific theoretical framework. Furthermore, one should avoid using multiple indicators that are highly correlated. The multicollinearity problems among the correlated indicators should be examined before building a measurement model with them. Finally, the construct validity of a measurement model should be examined carefully. The reliability of the model can be enhanced if more than two related indicators are used to reflect the construct that one has to be measured.

The following is an empirical example to show how a measurement model of health status, a latent construct, is established and validated, using confirmatory factor analysis (Figure 28):

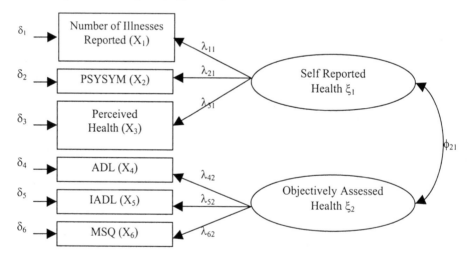

Figure 28. Confirmatory Factor Analysis of Health Status

94

The above path diagram shows the multi-dimensionality of health status. A one-way arrow between two variables indicates a postulated direct influence of one variable on another. A two-way arrow between two variables indicates that they are correlated. Arrows from the latent variable (ξ) to its indicator variables (X_i) are factor loadings or lambda coefficients (λ_{ij}), where "i" refers to the variable at the receiving end of the causal path, and "j" the latent variable that the arrow is coming from. The measurement errors are deltas (δs). The intercorrelation between the two constructs or latent variables is ϕ. For each variable or indicator, there is a structural equation for each one-way arrow linked with the latent variable: $X = f(\xi) + \delta$. There are six estimation equations for the measurement model of health status indicators.

$$X_1 = \lambda_{11} \xi_1 + \delta_{11} \quad \dots\dots\dots(6\text{-}2).$$
$$X_2 = \lambda_{21} \xi_1 + \delta_{22} \quad \dots\dots\dots (6\text{-}3).$$
$$X_3 = \lambda_{31} \xi_1 + \delta_{33} \quad \dots\dots\dots (6\text{-}4).$$
$$X_4 = \lambda_{42} \xi_2 + \delta_{44} \quad \dots\dots\dots (6\text{-}5).$$
$$X_5 = \lambda_{52} \xi_2 + \delta_{55} \quad \dots\dots\dots 6\text{-}6).$$
$$X_6 = \lambda_{62} \xi_2 + \delta_{66} \quad \dots\dots\dots (6\text{-}7).$$

The above equations can be summarized into a matrix form as follows:

$$
\begin{vmatrix} X_1 \\ X_2 \\ X_3 \\ X_4 \\ X_5 \\ X_6 \end{vmatrix}
=
\begin{vmatrix} \lambda_{11} & 0 \\ \lambda_{21} & 0 \\ \lambda_{31} & 0 \\ 0 & \lambda_{42} \\ 0 & \lambda_{52} \\ 0 & \lambda_{62} \end{vmatrix}
\begin{vmatrix} \xi_1 \\ \xi_2 \end{vmatrix}
+
\begin{vmatrix} \delta_{11} \\ \delta_{22} \\ \delta_{33} \\ \delta_{44} \\ \delta_{55} \\ \delta_{66} \end{vmatrix}
= \underset{\sim}{\Lambda}\underset{\sim}{\xi} + \underset{\sim}{\delta}.
$$

The following is a LISREL program for modeling health status for 694 elderly subjects.

```
CONFIRMATORY FACTOR ANALYSIS OF HEALTH STATUS INDICATORS
DA NI=6 NO=694 MA=KM
LA
'MSQ' 'PSYSYM' 'PERHEAL' 'ADL' 'IADL' 'NILL'
KM SY
```

```
1.000
 .096   1.000
 .061    .498   1.000
 .238    .233    .202   1.000
 .383    .293    .203    .470   1.000
 .010    .378    .236    .105    .114   1.000
SE
 2 3 4 5 1
MO NX=6 NK=2 PH=ST
LK
'SELF'  'ASSESS'
FR LX(1,1)  LX(2,1)  LX(3,1)  LX(4,2)  LX(5,2)  LX(6,2)
OU SE TV RS EF VA MI AM SS AD=OFF
```

APPLICATION: AN EXAMPLE

This example evaluates Hertzog's four-factor measurement model of depression (Ensel, 1986;Hertzog, 1988) through confirmatory factor analysis.

Data for this study were drawn from the National Health Examination Survey Epidemiologic Follow-up Study. The cross-sectional data were collected from a national probability sample of the United States civilian noninstitutionalized adult population, 25 years old and older. A subsample of 3,562 responses from the original sample was randomly selected to be analyzed.

The Center for Epidemiological Studies Depression Scale (CES_D) is a 20-item, self - reported scale designed to measure the prevalence of depressive symptoms in a community (Radloff, 1977). The scale comprises 20 items that are considered representative of four major components of depressive symptomatolgy: depressive affect, level of well-being, somatic symptoms, and interpersonal problems (Figure 29). Each of the items asks for the frequency with which a given symptom was experienced during the previous week, using a scale shown in Exhibit 1. A four-level frequency rating scale score (0-3) was used. In addition, scores of 7, 8 and 9 were used to indicate "inapplicable," "don't know," and "not ascertained," respectively. Scoring for positively worded items has been reversed so that high scores represent responses in the depressed range.

Exhibit 1. CES-D Depression Scale: The following questions are asked of each respondent. Items can be treated as observed indicators (X_1 to X_{20}).

I1. During the past week things that usually don't bother me bothered me.
I2. I did not feel like eating; my appetite was poor.
I3. I felt that I could not shake off the blues even with help from my family or friends.
I4. I felt that I was just as good as other people.
I5. I had trouble keeping my mind on what I was doing.
I6. I felt depressed.
I7. I felt that everything I did was an effort.
I8. I felt hopeful about the future.
I9. I thought my life had been a failure.
I10. I felt fearful.
I11. My sleep was restless.
I12. I was happy.
I13. I talked less than usual.
I14. I felt lonely.
I15. People were unfriendly.
I16. I enjoyed life.
I17. I had crying spells.
I18. I felt sad.
I19. I felt that people disliked me.
I20. I could not "get going."

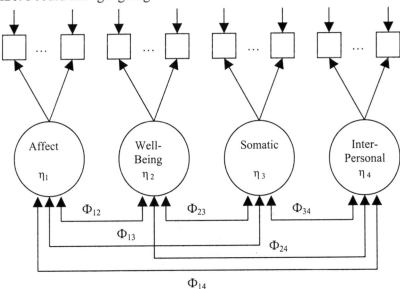

Figure 29. The Generic Measurement Model of Depression

97

Based upon an exploratory factor analysis and Hertzog's article, the list of items that are associated with four latent constructs or variables is as follows:

Affect: I1 I3 I6 I10 I14 **I17** I18
Somatic: I2 I5 **I7** I11 I13 I20
Well-being: I4 I8 I12 **I16**
Inter-personal: I9 **I15** I19

__ denotes the strongest factor loading in each factor.
The specification of this measurement model is presented in Exhibit 2.

Exhibit 2. Specification of the Proposed Model

		col 1	col 2	col 3	col 4			
X1		λ_{11}						δ_1
X2				λ_{23}				δ_2
X3		λ_{31}						δ_3
X4			λ_{42}					δ_4
X5				λ_{53}				δ_5
X6		λ_{61}						δ_6
X7				λ_{73}		ξ_1		δ_7
X8			λ_{82}			ξ_2		δ_8
X9	=				λ_{94}	ξ_3	+	δ_9
X10		λ_{10}				ξ_4		δ_{10}
X11				λ_{113}				δ_{11}
X12			λ_{122}					δ_{12}
X13								δ_{13}
X14		λ_{141}						δ_{14}
X15					λ_{154}			δ_{15}
X16			λ_{162}					δ_{16}
X17		λ_{171}						δ_{17}
X18		λ_{181}						δ_{18}
X19					λ_{194}			δ_{19}
X20				λ_{203}				δ_{20}

A LISREL program for the initial model is presented below.

```
DA NI=20 NO=3562 MA=KM
LA
'I1' 'I2' 'I3' 'I4' 'I5' 'I6' 'I7' 'I8' 'I9'
'I10' 'I11' 'I12' 'I13' 'I14''I15' 'I16' 'I17' 'I18' 'I19' 'I20'
KM SY
1. 000
0.316  1. 000
0.470   0.359 1.000
0.017 -0.029 0.008 1.000
0.349   0.271 0.416 -0.037 1.000
0.466   0.334 0.621 0.012 0.482 1.000
0.398   0.339 0.440 -0.034 0.413 0.509 1.000
0.099    0.061 0.146 0.415 0.065 0.154 0.098 1.000
0.267   0.233 0.384 0.072 0.319 0.408 0.323 0.108 1.000
0.319   0.254 0.412 0.033 0.371 0.475 0.360 0.104 0.455 1.000
0.280   0.275 0.349 0.009 0.301 0.376 0.335 0.090 0.271 0.329 1.000
0.220   0.135 0.274 0.322 0.179 0.311 0.215 0.437 0.222 0.236 0.195
1.000
0.277   0.260 0.322 0.034 0.289 0.334 0.295 0.085 0.281 0.282 0.256
0.114 1.00
0.323   0.257 0.465 0.025 0.358 0.548 0.375 0.128 0.385 0.413 0.335
0.343
1. 000
0.186   0.172 0.278 0.029 0.238 0.277 0.228 0.066 0.336 0.287 0.171
0.124 0.234
0.332   1. 000
0.199    0.126 0.235   0.340 0.152 0.257 0.181 0.421 0.190 0.184 0.155
0.636 0.145
0.209   0.096 1. 000
0.340   0.280 0.484   0.028 0.304 0.402 0.332 0.118 0.343 0.399 0.291
0.228 0.258
0.444   0.270 0.187   1. 000
0.411   0.286 0.543   0.036 0.385 0.612 0.429 0.139 0.406 0.469 0.368
0.299 0.323
0.557   0.301 0.238   0.598 1.000
0.227   0.168 0.318   0.059 0.268 0.344 0.252 0.098 0.406 0.332 0.208
0.163 0.247
0.361   0.486 0.151   0.352 0.384 1.000
0.323 0.276 0.399 0.003 0.410 0.431 0.524 0.073 0.315 0.343 0.370
0.181 0.317
0.380   0.227 0.154 0.371 408 0.288 1.000
SD
0.698 0.617 0.598 1.205 0.722 0.682 0.880 1.193 0.526 0.528 0.874
```

```
1.040 0.813
0.68 0.554 0.994 0.321 0.612 0.488 0.757
SE
1 3 6 10 14 17 18 2 5 7 11 13 20 4 8 12 16 9 15 19
MO NX=20 NK=4 PH=SY,FR
LK
'AFFECT' 'SOMATIC' 'WELBEING' 'INTPRSNL'
FR LX(1,1) LX(2,1) LX(3,1) LX(4,1) LX(5,1) LX(7,1) LX(8,2) LX(9,2)
LX(11,2
FR LX(12,2) LX(13,2) LX(14,3) LX(15,3) LX(16,3) LX(18,4) LX(20,4)
ST 1 LX(6,1) LX(10,2) LX(17,3) LX(19,4)
OU SE TV RS EF VA MI ADD=OFF
```

Exhibit 3. Specification of the Revised Model

$$
\begin{bmatrix}
X1 \\ X2 \\ X3 \\ X4 \\ X5 \\ X6 \\ X7 \\ X8 \\ X9 \\ X10 \\ X11 \\ X12 \\ X13 \\ X14 \\ X15 \\ X16 \\ X17 \\ X18 \\ X19 \\ X20
\end{bmatrix}
=
\begin{bmatrix}
 & & \lambda_{13} & \\
 & & \lambda_{23} & \\
\lambda_{31} & & \lambda_{33} & \\
 & \lambda_{42} & & \\
 & & \lambda_{53} & \\
\lambda_{61} & & & \\
 & & \lambda_{73} & \\
 & \lambda_{82} & & \\
\lambda_{91} & & & \\
\lambda_{10} & & & \\
 & & \lambda_{113} & \\
 & \lambda_{122} & & \\
 & & & \\
\lambda_{141} & & & \\
 & & & \lambda_{154} \\
 & \lambda_{162} & & \\
\lambda_{171} & & & \\
\lambda_{181} & & & \\
 & & & \lambda_{194} \\
 & & \lambda_{203} &
\end{bmatrix}
\begin{bmatrix}
\xi_1 \\ \xi_2 \\ \xi_3 \\ \xi_4
\end{bmatrix}
+
\begin{bmatrix}
\delta_1 \\ \delta_2 \\ \delta_3 \\ \delta_4 \\ \delta_5 \\ \delta_6 \\ \delta_7 \\ \delta_8 \\ \delta_9 \\ \delta_{10} \\ \delta_{11} \\ \delta_{12} \\ \delta_{13} \\ \delta_{14} \\ \delta_{15} \\ \delta_{16} \\ \delta_{17} \\ \delta_{18} \\ \delta_{19} \\ \delta_{20}
\end{bmatrix}
$$

LISREL Program: the Revised Model

```
DA NI=20 NO=3562 MA=KM
LA
'I1' 'I2' 'I3' 'I4' 'I5' 'I6' 'I7' 'I8' 'I9' 'I10' 'Ill'
'I12' 'I13'
'I14' 'I15' 'I16' 'I17' 'I18' 'I19' 'I20'
KM SY
1. 000
0.316 1.000
0.470 0.359 1.000
0.017 -0.029 0.008 1.000
0.349 0.271 0.416 -0.037 1.000
0.466 0.334 0.621 0.012 0.482 1.000
0.398 0.339 0.440 -0.034 0.413 0.509 1.000
0.099 0.061 0.146 0.415 0.065 0.154 0.098 1.000
0.267 0.233 0.384 0.072 0.319 0.408 0.323 0.108 1.000
0.319 0.254 0.412 0.033 0.371 0.475 0.360 0.104 0.455
1.000
0.280 0.275 0.349 0.009 0.301 0.376 0.335 0.090 0.271
0.329 1.000
0.220 0.135 0.274 0.322 0.179 0.311 0.215 0.437 0.222
0.236 0.195 1.000
0.277 0.260 0.322 0.034 0.289 0.334 0.295 0.085 0.281
0.282 0.256 0.114 1.00
0.323 0.257 0.465 0.025 0.358 0.548 0.375 0.128 0.385
0.413 0.335 0.343 0.340
1. 000
0.186 0.172 0.278 0.029 0.238 0.277 0.228 0.066 0.336
0.287 0.171 0.124 0.243
0.332 1.000
0.199 0.126 0.235 0.340 0.152 0.257 0.181 0.421 0.190
0.184 0.155 0.636 0.145
0.209 0.096 1.000
0.340 0.280 0.484 0.028 0.304 0.402 0.332 0.118 0.343
0.399 0.291 0.228 0.258
0.444 0.270 0.187 1.000
0.411 0.286 0.543 0.036 0.385 0.612 0.429 0.139 0.406
0.469 0.368 0.299 0.323
0.557 0.301 0.238 0.598 1.000
0.227 0.168 0.318 0.059 0.268 0.344 0.252 0.098 0.406
0.332 0.208 0.163 0.247
0.361 0.486 0.151 0.352 0.384 1.000
0.323 0.276 0.399 0.003 0.410 0.431 0.524 0.073 0.315
```

```
0.343 0.370 0.181 0.317 0.380 0.227 0.154 0.371 0.408
0.288 1.000
SD
0.698 0.617 0.598 1.205 0.722 0.682 0.880 1.193 0.526
0.528 0.874 1.040 0.813
0.68 0.554 0.994 0.321 0.612 0.488 0.757
SE
1 3 6 9 10 14 17 18 2 5 7 11 13 20 4 8 12 16 /
MO NX=18 NK=3 PH=SY,FR
LK
'AFFECT' 'SOMATIC' 'WELBEING'
FR LX(1,1) LX(2,1) LX(3,1) LX(4,1) LX(5,1) LX(7,1)
LX(8,1) LX(9,2) LX(10,2)
FR LX(12,2) LX(13,2) LX(14,3) LX(15,3) LX(16,3) LX(17,3)
ST 1 LX(6,1) LX(11,2) LX(18,3)
OU SE TV RS EF VA MI ADD=OFF
```

The proposed measurement model of depression is depicted in Figure 30. Factor equations and the matrix form for this model can be found in Exhibit 1. It is found that in order to identify the CFA model, there are 40 unknowns in the factor equations that must be solved for, and 210 known factors from the variance--covariance matrix to be used in solving for the unknowns. This indicates that the proposed model is over-identified, so a confirmatory factor analysis can be carried out.

All of the Φ's (Phi's), which represent intercorrelations between the latent variables, are also statistically significant. The strongest intercorrelation occurs between the AFFECT and SOMATIC constructs. Positive affect correlates rather poorly with all of the other constructs, while INTERPERSONAL correlates moderately to strongly with AFFECT and SOMATIC.

All the measurement errors included in the four-factor model are statistically significant. This may indicate that the CES - D's items are not perfectly measured for reflecting the latent constructs of the scale. In addition, the extremely small determinant for the model indicates strong multicollinearity among the scale items. This suggests that some items could probably be dropped from the scale without compromising model fit.

The reliability of the individual items is indicated by their squared multiple correlations. These tend to be low or moderately low for most of the items. Overall, the items measuring depressed affect seem to be more reliable than those of other constructs. I6 is the most reliable for depressed affect (AFFECT), I12 for positive or well being affect (WELL-BEING), I7 for

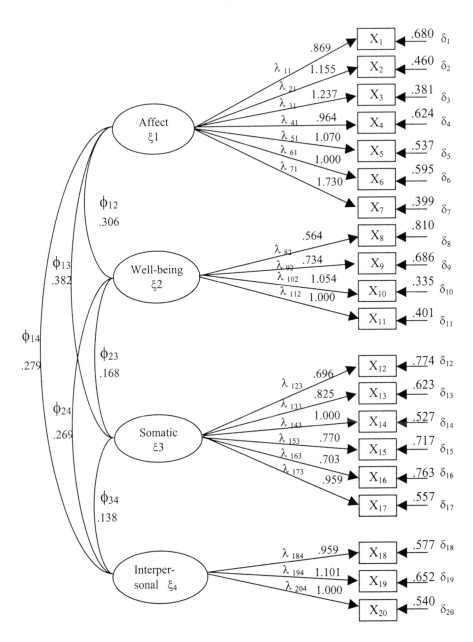

Figure 30. A Proposed Measurement Model

103

somatic (SOMATIC) and I19 for interpersonal (INTPRSNL).

Table 7 summarizes the measures of goodness of fit. Although the coefficient of determination indicates high reliability for the measurement model, the very large and significant chi - square and the large likelihood ratio indicate that the model is poorly fit for the sample data. The goodness of fit (GOF) index is also marginal. Further evidence of a poor fit is seen in the examination of the normalized residuals. A number of these residuals are very large, and a Q - plot of the residuals shows the slope of the line they create to be less than 1, indicating a poor fit. This suggests that the model could probably be improved for the current data.

Table 7. Goodness of Fit Measures for Proposed and Revised Models of Depression Measurement

Index	Proposed	Revised
Chi - Square (χ^2)	1978.33	1048.99
Degrees of Freedom (df)	164	106
Probability (p)	0.00	.00
Likelihood Ratio (χ^2/df)	12.0629	9.896
GOF Index	.946	.967
Adjusted GOF Index	.931	.952
Root Mean Square Ratio	.043	.025

Note: GOF is the goodness of fit.

Figure 31 illustrates the revised measurement model for depression. Factor equations and the matrix form for this model can be found in Exhibit 3. The model was revised on the basis of the theoretical rationale. Consideration was given to whether the interpersonal relations construct should logically be linked to depression. Although problems in interpersonal relations are likely to be a consequence of depression, the construct does not necessarily correlate with depression. For this reason, the entire construct was deleted from the revised model. Item 9 was retained and linked with the depressed affect construct. This was supported by the fact that Hertzog (1988) found item 9 to load on this construct during exploratory factor analysis of two separate samples.

The revised measurement model is found to be over-identified, with 36 unknowns to be solved for and 171 knowns to use in solving for the unknowns. A confirmatory factor analysis of the three-factor model was therefore carried out. The phi's in the revised model are all statistically significant and are similar to those in the proposed model. The deltas in the revised model also remain largely unchanged and statistically significant. The five intercorrelations of the measurement errors that were added to this model are also statistically significant. However, the X^2 remains large. The decrease in the X^2 from one model to another can be tested for statistical significance. In this case the decrease in chi - square is 930, with df = 164 - 106 = 59. At the .05 alpha level, this indicates that the fit of the model has improved moderately. By increasing the number of intercorrelations between the measurement errors, the fit of the model could probably be improved further. However, there is a point beyond which all improvements have to be considered purely empirical, and not in the

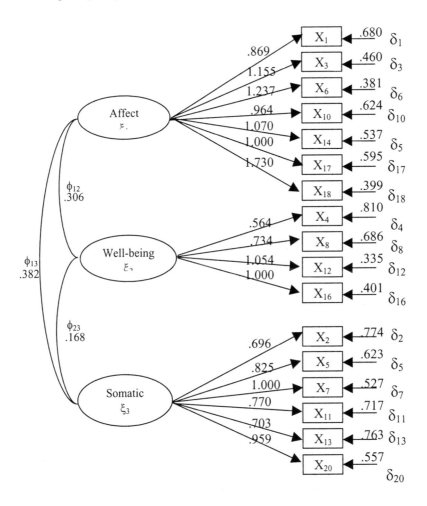

Figure 31. Revised Measurement Model

105

best interests of the theoretical model. For this reason, only the strongest five intercorrelations were added to the revised model.

In conclusion: dropping the interpersonal relational construct improved the model fit moderately for the study sample, although the large X^2 and the likelihood ratio indicate that the fit is still not as good as it could be. However, according to Hertzog (1988), a model can still be considered useful even though the likelihood ratio chi - square test is statistically significant. With large sample sizes, the likelihood ratio test is very powerful and can determine if the model accurately represents the correlations matrix. In addition, violation of the assumption of multivariate normality of the data can result in inflated likelihood ratios. In view of the fact that the current sample is large and violates the normality assumption, it is reasonable to conclude that the revised model may actually fit the data better than the chi - square statistics indicate.

Although the revised measurement model may not exhibit a strong fit for the data, this does not necessarily invalidate the model. It simply indicates that the current data does not confirm this model. Replication with new data could resolve the issue of the model' s validity. A second-order factor (e.g., general psychological well-being) may exist among the three dimensions of psychological well-being, as well.

AN APPLICATION OF CONFIRMATORY FACTOR ANALYSIS: SF-36

Following the seven steps in causal analysis outlined in Chapter 2, an analysis of the SF-36 Health Status Measurement is presented below.

1. Identification and Specification of the Study Problem

Health is an elusive concept, in that the word conveys numerous meanings depending on how it is used. Health care providers continue to search for the perfect tool that will allow a rapid and accurate assessment of an individual's health. The Short-Form 36 (SF-36) has shown great promise in that direction. The SF-36 was developed as part of the medical outcomes study (Ware and Sherbourne, 1992). The instrument has been used in numerous studies and translated into different languages to test its ability to accurately measure perception of health (www.sf-36.com; Hays et al., 1994, Hays, Morales, and

Reise, 2000; McHorney, Ware, and Raczek, 1993; McHorney et al., 1994). Despite wide acceptance and utilization of the tool, the underlying constructs of Sf-36 remain to be validated.

The SF-36 was developed to provide generic measures of health from the patient's point of view (Ware et al., 1993; Ware and Sherbourne, 1992); Ware et al., 1995). The ability of the instrument to measure health also makes it a good tool for measuring outcomes of care and treatment. The majority of efforts to determine the reliability and validity of the SF-36 have been exploratory and conducted across a wide variety of settings. It is unclear how well aggregated indices measured by the SF-36 represent major health concepts in terms of the construct validity of subindices (Pai & Wan, 1997). Does the tool measure two or three constructs, or merely one overarching construct of health status?

The purpose of this illustration is to evaluate the construct validity of the SF-36, using confirmatory factor analysis. Drawing on the results from exploratory factor analysis conducted by the medical outcomes study group (MOS), cross-sectional data are employed to examine the measurement integrity of a two-factor health status model measured by the SF-36 and to revise the initial model for a better model fit. Confirmatory factor analysis provides an appropriate factor-analytic strategy to extract the health constructs of the SF-36.

2. Selection of an Informed Theoretical Framework

The World Health Organization (WHO) defines health as "physical, mental, and social well-being, and not merely the absence of disease and infirmity" (World Health Organization, 1958, p.459). It is the operationalization of these three concepts that remains inconsistent in the health care community. Health professionals consider that health is a state of biopsychosocial being in a continuum of illness to wellness. The goal of health promotion and disease prevention activities is to help individuals achieve optimal levels of wellness. This perspective suggests that there are three related concepts involved in the determination of health. The foundation of the SF-36 is long-form measures that were constructed to measure two of the major dimensions of health: physical health and mental health (Reed, 1998). The developers of the tool do not suggest that the tool adequately measures social wellness. Only one item measures social functioning, despite the importance of this construct to overall health status.

The 36-item, Short-Form Health Survey (SF-36) was developed by Dr. John Ware and his Medical Outcomes Study (MOS) colleagues to provide generic measures of health status and outcomes from the patient's point of view. Such generic measures are not specific to age, disease, or treatment. The SF-36 is practical because it is shorter than many other such surveys and can be self-administered. By constructing scales from more efficient items, the SF-36 attempts to reduce the burden on the respondent without bringing measurement precision below the critical level. The SF-36 has been used in various health care settings with different patient groups. Only a few are mentioned below. The SF-36 has been shown to be useful for gathering new information for physicians utilizing new treatments for patients with diabetes. The SF-36 was found to offer useful information about general health status, role functioning, and well-being, particularly for outpatient dialysis recipients.

The SF-36 is used increasingly to measure outcomes for patients who need rehabilitative care, such as total hip replacement, total knee replacement, motor neuron disease, and shoulder surgery.

SF-36 Measures

The SF-36 is a widely used measure of health care outcomes from a patient's perspective. The SF-36 measures nine health concepts: (1) general health perceptions (general health); (2) health transition; (3) limitations on physical activities because of health problems (physical functioning); (4) limitations on usual role activities because of physical health problems (role-physical); (5) limitations on usual role activities because of emotional problems (role-emotional); (6) limitations on social activities because of physical or emotional problems (social functioning); (7) bodily pain; (8) psychological distress or well-being (mental health); and (9) energy and fatigue (vitality). Scales are scored using Likert's method of summated ratings, which assumes that the distributions of responses to items within the same scale and item variances are roughly equal. Each item is also assumed to have a substantial linear relationship with the score for its scale, that is, item internal consistency. The use of each item to score only one scale assumes substantial item discriminant validity, that is, each item clearly measuring one health concept more than other health concepts. When these assumptions are well satisfied, items in the same scale can be scored without standardization and can be simply summed. Results from SF-36 tests of scaling assumptions strongly support the use of summated ratings to compute SF-36 subscales.

SF-36 subscales are scored so that a higher score indicates a better health state. In a 1992 article, Ware and Sherbourne present information about SF-36 health status scales and the interpretation of low and high scores. Transforming each raw scale score to a 0 to 100 scale is strongly recommended. Estimates of score reliability for eight SF-36 subscales (internal consistency reliability, test-retest reliability, or alternate-form reliability) have been reported in 15 studies. All estimates show high reliability, ranging from 0.6 to 0.96. Physical functioning tends to have the highest internal consistency reliability, probably because more items are used to measure this construct.

Criterion validity and precision have been well established for the eight scales found of the SF-36 (Ware & Sherbourne, 1992). This instrument is used routinely in some settings and regularly in research studies to look at the outcomes of a variety of groups of patients. Most recently, two articles examine how well 6-month mortality for post-coronary artery bypass grafting correlates with poor scores on health before surgery (Rumsfeld, et al., 1999; Sahin, Wan, and Sahin, 1999).

3. Quantification of the Study Variables

The subjects of this study were drawn from a sample of enrollees in a managed care program. Data were collected in 1994 during the pretest phase of a large research project for comprehensive assessment of cost, satisfaction, and quality among different service products. Inclusion criteria required that a managed care policyholder be an employee since January 1993, have continuous group coverage since January 1993, and be at least 18 years old but less than 65 years old in July, 1994. All randomly sampled subjects were selected from the membership and claims information system of the company. A self-administered questionnaire containing the SF-36 survey was sent to the study sample. The final sample consisted of 2,375 responses.

4. Specification in Analytical Modeling

The computer program AMOS 4 was used to examine how well the eight aggregate indices (X_1-X_8) obtained from the SF-36 represent two common latent variables, physical status (ζ_1) and emotional status (ζ_2). Both latent variables are to be correlated with each other. Similarly, correlated measurement errors (δs) among the indicators are assumed. The initial generic model presented in Figure 6-8 is based partly on the results from the exploratory factor analysis conducted by the MOS. The revised model shown in Figure 32 allowed measurement errors to be correlated with each other. In addition, aggregate indices were allowed to be loaded onto more than one latent construct if they were postulated to be correlated.

5. Selection of the Intervention Design

This is not an experimental study. Intervention design is not applicable.

6. Confirmatory Factor Analysis

Table 8 shows the demographics of the study sample. The average age of respondents in 1990 was 49. Fifty-seven percent of the sample were female and 84% of the sample were white. The author is unable to determine the other categories of race, because the database did not come with a coding key.

More than half of the respondents were high school graduates. Table 9 displays a correlation matrix and the standard deviations for the eight aggregate indices. All indices were statistically significantly correlated at the p = .000 level.

The initial model in Figure 33 was specified to be:

$$
\begin{vmatrix} X1 \\ X2 \\ X3 \\ X4 \\ X5 \\ X6 \\ X7 \\ X8 \end{vmatrix}
=
\begin{vmatrix} \lambda 11 & 0 \\ \lambda 21 & 0 \\ \lambda 31 & 0 \\ \lambda 41 & 0 \\ \lambda 51 & \lambda 52 \\ \lambda 61 & \lambda 62 \\ 0 & \lambda 72 \\ 0 & \lambda 82 \end{vmatrix}
*
\begin{vmatrix} \xi 1 \\ \xi 2 \end{vmatrix}
+
\begin{vmatrix} \delta 1 \\ \delta 2 \\ \delta 3 \\ \delta 4 \\ \delta 5 \\ \delta 6 \\ \delta 7 \\ \delta 8 \end{vmatrix}
$$

110

Table 8. Demographic Characteristics of the Study Sample

Variable	N (%)	Min/Max	Mean
Gender			
Male	1015 (42.7)		
Female	1356 (57.1)		
Race			
White	2000 (84.2)		
Black	208 (8.8)		
Other	162 (7.0)		
Age	2366 (99.6)	18/94	49
Education in years	2343 (98.6)	1/17	12.8

Table 9. Correlation Matrix of Eight Aggregate Indices

	PFI	ROLEP	PAIN	GHP	VITALITY	SOCIAL	ROLEE	MH
PFI	1							
ROLEP	0.67	1						
PAIN	0.549	0.	1					
GHP	0.582	0.573	0.582	1				
VITALITY	0.494	0.540	0.553	0.608	1			
SOCIAL	0.467	0.547	0.519	0.494	0.527	1		
ROLEE	0.336	0.453	0.350	.0390	0.448	0.540	1	
MH	0.287	0.348	0.394	0.463	0.622	0.559	0.548	1
Std. Dev.	24.9	36.61	24.4	20.92	21.51	23.35	34.34	18.31

Note: PFI = physical functioning (X_1), ROLEP = role-physical (X_2), PAIN = bodily pain (X_3), GHP = general health perception (X_4), VITALITY = vitality (X_5), SOCIAL = social functioning (X_6), ROLEE = role-emotional (X_7), MH = mental health (X_8).

Table 10 shows the parameter estimates for the initial and revised models. The initial model was revised and measurement errors were allowed to be correlated if their modification indices were greater than 50 and the correlation made theoretical sense. The regression weight of 1 was assigned to ROLEP and MH in both models. The initial model proposes that vitality and social functioning are indices of physical status and emotional status, while correlating the two latent constructs. The correlation of the latent constructs was 0.623 in the initial model and 0.665 in the revised model. Both were beneath the recommended cutoff of 0.7. Based on the WHO definition and nursing theory, both constructs must be measured in order to ascertain health status. Figures 32 and 33 show the two models with their parameter estimates.

Table 10. Parameter Estimates for the Initial and Revised Models

Lambda	Initial Model		Revised Model	
	Physical Status	Emotional Status	Physical Status	Emotional Status
PFI	0.761		0.721	
ROLEP	0.816		0.810	
PAIN	0.765		0.764	
GHP	0.758		0.792	
VITALITY	0.421	0.444	0.546	0.249
SOCIAL	0.355	0.467	0.288	0.549
ROLEE		0.681		0.724
MH		0.831		0.760
Delta Values				
PFI	0.42		0.48	
ROLEP	0.33		0.34	
PAIN	0.41		0.42	
GHP	0.43		0.37	
VITALITY	0.39		0.45	
SOCIAL	0.45		0.41	
ROLEE	0.54		0.48	
MH	0.31		0.42	
δ_{85}			0.350	
δ_{82}			-0.080	
δ_{21}			0.204	
δ_{42}			-0.186	
δ_{72}			0.166	

Note: PFI = physical functioning, ROLEP = role-physical, PAIN = bodily pain,
GHP = general health perception, VITALITY = vitality, SOCIAL = social functioning,
ROLEE = role-emotional, MH = mental health.

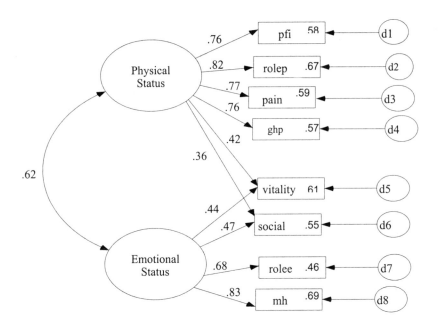

Figure 32. Model Without Correlated Measurement Errors

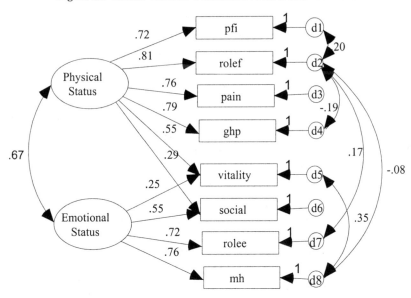

Figure 33. Model With Correlated Measurement Errors

Table 11 compares the goodness of fit measures between the initial and revised models. The chi-square and p values of the initial and revised model indicate that the model remains not optimally fit. This does not mean that the model was poorly specified, but only that it does not fit optimally with these data. The model is recursive and overspecified, which can be seen by the chi-square and the likelihood ratio. However, the revised model demonstrates improvement in all of the goodness of fit statistics.

Table 11. Goodness of Fit Statistics

Index	Initial Model	Revised Model
Chi-square	529.43	154.73
Degrees of Freedom	17	12
Probability	0.000	0.000
Likelihood Ratio ($\chi2$ / df)	31.14	12.89
GOF Index	0.942	0.984
Adjusted GFI	0.878	0.953
RMSEA	0.113	0.071
Hoelter N @ .05	124	323

7.   Establishment of Causality

The revised model demonstrated that the reliability coefficients do a good job explaining the variance of each indicator for a given theoretical construct. The delta values indicate that much more variance needs to be explained. This may be a result of how the model was constructed. There may also be one overarching construct that the two latent constructs in this model are tied to. This line of thinking corroborates what other studies with various patient groups have discovered, that the SF-36 is a good measure of overall health status. There are other general issues related to using the SF-36 to measure health status. Important health concepts are not represented in the SF-36. Some of those omitted are under study: health distress, family functioning, sexual functioning, cognitive functioning, and sleep disorders (Ware & Sherbourne, 1992). A tool that incorporates the missing pieces would round out the SF-36 nicely (perhaps to an SF-40). This instrument also does not assess social functioning well enough to have it represented by its own construct. This lack is contrary to many definitions of health and wellness and leaves a significant gap in measuring health status. Finally, it should be noted that short form measures typically have at least two types of problems, floor effects and ceiling effects. Subsequent studies with the SF-36 have demonstrated that floor effects are rare (Ware & Sherbourne, 1992; Hays, 2000). However, when it is used for studies of severely ill populations, it may be wise to add an additional series of tests to represent the extreme low end of the continuum (Ware & Sherbourne, 1992). The SF-36 survey instrument has been proven a useful tool. Apparently, it remains to be determined exactly which latent constructs the tool measures best. The perfect measurement tool is an elusive concept. While it is tempting to want to use a good tool for many purposes, caution is advised. The SF-36 has been supported for group-level analyses. Further research should be done to evaluate the appropriateness of the tool for monitoring outcomes in individual patients (Sahin, Wan, and Sahin, 1999). Using item response theory, the construction of self-reported health and outcome measures can be significantly improved for this tool's validity and reliability (Hays, Morales and Reise, 2000).

CONCLUSION

Health status instruments have been used for various purposes, including health policy evaluations, monitoring the health of general populations, and designing systems to monitor and improve health care outcomes (Chern, Wan, and Pyles, 2000; McHorney et al., 1994). The validity, reliability and applicability of the health status measurement should be fully demonstrated before it is used to perform program assessment and evaluation in a study population. Furthermore, the stability of the measurement model should be confirmed with panel data.

REFERENCES

Bollen, K.A. (1989). *Structural Equations with Latent Variables*. New York, NY: John Wiley & Sons.

Chern, J.Y., Wan, T.T.H., Plyes, M. (2000). The stability of health status measurement (SF-36) in a working population. *Journal of Outcome Measurement* 4(1): 461-481.

Ensel, W. (1986). Measuring depression: The CES - D scale. In N. Lin, A. Dean and W. Ensel (eds.) *Social, Support, Life Events and Depression*. New York: Academic Press.

Hays, R.D., Marshall, G.N., Wang, E.Y., Sherbourne, C.D. (1994). Four-year cross-lagged associations between physical and mental health in the Medical Outcomes Study. *Journal of Consulting Clinical Psychology* 62(3): 441-449.

Hays, R.D., Morales, L.S., Reise, S.P. (2000). Item response theory and health outcomes measurement in the 21st century. *Medical Care* 38(9 Suppl): II28-142.

Hertzog, C. (1988). Using confirmatory factor analysis for scale development and validation. In Lawton, P.M. & Herzog, A.R. (Eds.), *SpecialRresearch Methods for Gerontology* (pp. 281-306). New York: Baywood Press.

Jöreskog, K.G., Sörbom, D. (1984). *LISREL User's Guide*. Mooresville, IN.: Scientific Software, Inc.

Jöreskog, K. G., Sörbom, D. (1979). *Advances in Factor Analysis and Structural Equation Models*. Cambridge, MA: Abt.

Long, J. S. (1983). *Confirmatory Factor Analysis*. Newbury Park, CA: Sage Publications.

McHorney, C.A., Ware, J.E., Raczek, A.E. (1993). The MOS 36-item short form health survey (SF-36): II. psychometric and clinical tests of validity in measuring physical and mental health constructs. *Medical Care* 31(3): 247-263.

McHorney, C.A., et al. (1994). The MOS 36-item short-form health survey (SF-36): III. Tests of data quality, scaling assumptions, and reliability across diverse patient groups. *Medical Care* 32 (1): 40-66.

Pai, C.W., Wan, T.T.H. (1997). Confirmatory analysis of health outcome indicators. *Journal of Rehabilitation Outcomes Measurement* 1 (2): 48-59.

Radloff, L.S. (1977). The CES-D scale: A self-report depressive scale for research in the general population. *Journal of Applied Psychological Measurement* 1: 385-401.

Reed, P. (1998). Medical outcome study short form 36: Testing and cross-validating a second order factorial structure for health system employees. *Health Services Research* 33(5): 1363-1380.

Rumsfeld, J.S., MacWhinney, S., McCarthy, M., Shroyer, A., VillaNueva, C., O'Brien, M., Moritz, T., Henderson, W., Grover, F., Sethi, G., Hammermeister, K. (1999). Health-related quality of life as a predictor of mortality following coronary artery bypass graft surgery. Participants of the Department of Veterans Affairs Cooperative Study Group on Processes, Structures, and Outcomes of Care in Cardiac Surgery. *Journal of American Medical Association,* 281(14): 1298-1303.

Sahin, I., Wan, T.T.H., Sahin, B. (1999). The determinants of CABG patients' outcomes. *Health Care Management Science* 2: 215-222.

Ware, J.E., et al. (1993). *SF-36 Health Survey: Manual and Interpretation Guide*. Boston, Massachusetts: The Health Institute, New England Medical Center, 1993.

Ware, J. W., Sherbourne, C.D. (1992). The MOS 36-Item Short-Form Health Survey (SF-36): I. Conceptual Framework and Item Selection. *Medical Care* 30 (6): 473-483.

Ware, J.E., et al. (1995). Comparison of methods for the scoring and statistical analysis of SF-36 health profiles and summary measures: Summary of results from the medical outcomes study. *Medical Care* 33 (4 suppl.): AS264-AS279.

Ware, J. (2000). The SF-36 health survey. http://www.sf36.com/general/sf36.html, 3/4/2000.

World Health Organization. (1958). *The First Ten Years of the World Health Organization.* Geneva: World Health Organization.

CHAPTER 7

STRUCTURAL EQUATIONS FOR DIRECTLY OBSERVED VARIABLES: RECURSIVE AND NON-RECURSIVE MODELS

This chapter presents the structural equations for directly observed variables. The analytical approach is very similar to path analysis, described in Chapter 3. In LISREL modeling of structural equations with observed variables, path analysis can be further specified with the recursive model (X affects Y only) and the non-recursive model (Y_1 affects Y_2 and Y_2 also affects Y_1). For example, if one examines the relationship between inpatient care (Y_1) and outpatient care (Y_2) services used by patients, an investigator can assume that a reciprocal relationship exists between these two types of service use (i.e. that an increase in inpatient care will be associated with an increase in outpatient care, and vice versa).

Structural equation modeling can be used to formulate numerous analytical models: 1) regression model, 2) autoregressive model, 3) recursive model, 4) non-recursive model, 5) panel model with stability (over time), 6) models with direct and indirect effects, and 7) econometric model.

A GENERIC STRUCTURAL EQUATION MODEL WITHOUT LATENT VARIABLES

A generic statistical model is stated in Equation 7-1 with only observed variables, and no latent constructs included.

$$\underset{\sim}{Y} = B \underset{\sim}{Y} + \Gamma \underset{\sim}{X} + \underset{\sim}{\zeta} \ . \qquad (7.1)$$

This model has the following assumptions:
1) Set Y is a set of endogenous variables and X is a set of exogenous variables. Both X and Y variables are perfectly measured without errors. ($\Lambda_x = I$, $\Lambda_y = I$.)
2) No measurement errors ($\theta_\varepsilon = 0$, $\theta_\delta = 0$) are observed for X and Y.

3) Intercorrelations (ϕ) among the exogenous variables are to be considered.

The parameter matrix in the model for an endogenous variable that affects another endogenous variable is represented by B, and for an exogenous variable that affects another endogenous variable is represented by Γ. The intercorrelation between exogenous variables is represented by ϕ. The three matrices are: $\underset{\sim}{B}, \underset{\sim}{\Gamma}, \underset{\sim}{\phi}$.

$E(\underset{\sim}{Z}) = 0$.

Covariance matrices include $\Phi = \text{Cov}(\underset{\sim}{X})$, $\Phi = \text{Cov}(\underset{\sim}{\zeta})$.

Reduced form: $\underset{\sim}{y} = (I - \underset{\sim}{B})\underset{\sim}{X} + (\underset{\sim}{I} - \underset{\sim}{B})$.

Covariance structure:

$$\Sigma = \text{Cov(YX)} = \left| \begin{array}{l} (\underset{\sim}{I} - \underset{\sim}{B})^{-1} \underset{\sim}{\Gamma} \Phi \underset{\sim}{\Gamma}'(\underset{\sim}{I} - \underset{\sim}{B})^{r-1} + (\underset{\sim}{I} - \underset{\sim}{B})^{-1} \Phi (\underset{\sim}{I} - \underset{\sim}{B})^{-1} \\ \Phi \underset{\sim}{\Gamma}'(\underset{\sim}{I} - \underset{\sim}{B})^{r-1} \end{array} \right|$$

To construct a single regression equation or simple path model, one has to be familiar with some symbols used in constructing path diagrams.

| X |

An observed quantitative variable X

A latent variable

| Y |

An observed quantitative variable Y

A direct causal effect of X on Y with a random disturbance term Z:

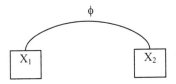

X_1 and X_2 may be correlated $\emptyset = \text{cov}(X_1, X_2)$

Let us assume that Y is a function of three predictor variables in Figure 34.

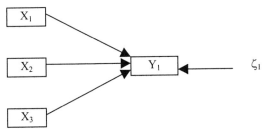

Figure 34. A Recursive Model

$$Y = X_1\gamma_{11} + X_2\gamma_{12} + X_3\gamma_{13} + \zeta_1. \qquad (7.2)$$

Parameter matrices are:

$B = 0 = (0)$ .... X does not affect Y.

$\Gamma = (\gamma_{11}, \gamma_{12}, \gamma_{13})$ .... X affects Y.

$\Phi = \text{Cov}(x)$, unconstrained, random or fixed.

The model is just identified (df = 0). Scientifically, the just identified model is not very interesting. Therefore, the model could be revised to generate an over-identified model by setting constraints on the parameters estimated.

A Model with Two Endogenous Variables

Several cases can be considered to determine the model specifications.

1) Y_1 and Y_2 are correlated?

2) Y_1 and Y_2 are correlated only because they have other causes that are correlated; or for given values of X_1, X_2, X_3, the variables Y_1 and Y_2 are uncorrelated; or the partial correlation between Y_1 and Y_2 is zero. Two disturbance terms or residuals (ζ_1 and ζ_2) of the equations are correlated (Figure 35).

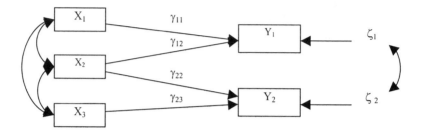

Figure 35. Two Unrelated Endogenous Variables With Common Causes

3) Y_1 affects Y_2 with different predictor variables (Figure 36). Before one can demonstrate this case, one must rule out cases 1 and 2.

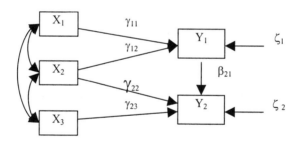

Figure 36. A Path Model with Y1 Affecting Y2

4) Y_1 and Y_2 depend on each other; a reciprocal causation exists (Figure 37). Before one can demonstrate this case, one must rule out cases 1, 2, and 3.

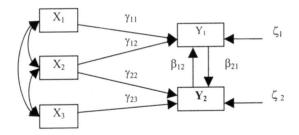

Figure 37. A Path Model with A Reciprocal Causation

5) Correlated Residual Terms: Suppose cases 1 and 2 do not hold and one does not believe they are causally related, or one does not want to commit that to 3 or 4. Then one can leave the correlation partly unexplained by assuming a shared common variance exists between the two residuals (ζ_1 and ζ_2) in Figure 38.

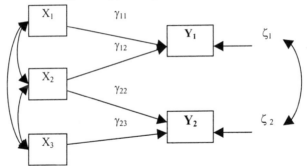

Figure 38. Correlated Residuals

SPECIFICATION OF PATH MODELS: EXAMPLES

Path Analysis of Two Unrelated Endogenous Variables

Let us assume that patient satisfaction (Y_1) and nursing staff satisfaction (Y_2) are unrelated in Figure 39, but they are influenced by the volume of services provided (X_1) and by nurses' work load (X_2). The causal link between X and Y is a path coefficient or gamma (γ). The residuals are assumed to be correlated because a shared common variance may be specified, but it is not related to X_1 and X_2.

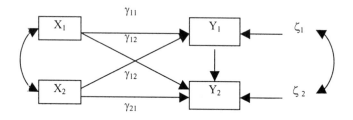

Figure 39. A Path Model with Correlated Residuals

Structural equations are:
$$Y_1 = \gamma_{11} X_1 + \gamma_{12} X_2 + \zeta_1.$$
$$Y_2 = \gamma_{21} X_1 + \gamma_{22} X_2 + \zeta_2.$$

Parameter matrices
$$\underset{\sim}{B} = 0.$$
2×2

$$\underset{\sim}{\Gamma} = \begin{vmatrix} \gamma_{11} & \gamma_{12} \\ \gamma_{21} & \gamma_{22} \end{vmatrix}$$
2×2

$\Phi = \text{Cov}(\underset{\sim}{X})$, unconstrained, random or fixed.
2×2

$$\underset{\sim}{\psi} = \begin{vmatrix} \psi_{11} & \psi_{12} \\ \psi_{21} & \psi_{22} \end{vmatrix}$$
2×2

The model is a just identified model (df = 0).

A Complete Recursive Model

Let us assume that the structural relationships among three dimensions of health functioning are as follows: 1) the physical functioning level (Y1) affects the mental functioning level (Y2); and 2) both physical and mental functioning levels affect the social functioning level. Three predictor variables (age [X_1], vitality [X_2], and socioeconomic status [X_3]) directly affect each of the three functioning dimensions in Figure 40. Residuals of the equations are not correlated to each other.

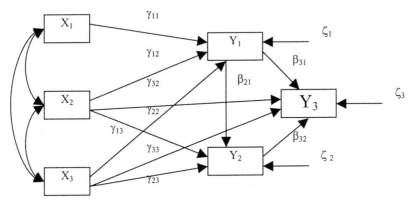

Figure 40. A Recursive Path Model

Structural equations are:

$Y_1 = \gamma_{11} X_1 + \gamma_{12} X_2 + \zeta_1.$

$Y_2 = \gamma_{23} X_3 + \gamma_{22} X_2 + \beta_{21} Y_1 + \zeta_2.$

$Y_3 = \gamma_{32} X_2 + \gamma_{33} X_3 + \beta_{31} Y_1 + \beta_{32} Y_2 + \zeta_3.$

Parameter matrices are:

$$
\underset{\sim}{B} = \begin{vmatrix} 0 & 0 \\ \beta_{21} & 0 \\ \beta_{31} & \beta_{32} \end{vmatrix}
\quad
\Gamma = \begin{vmatrix} \gamma_{11} & \gamma_{12} & 0 \\ 0 & \gamma_{22} & \gamma_{23} \\ 0 & \gamma_{32} & \gamma_{33} \end{vmatrix}
$$

$\Phi = \text{Cov} (\underset{\sim}{X})$ Unconstrained, fixed or random.
3×3

$$\underset{\sim}{\psi} = \begin{vmatrix} \psi_{11} & 0 & 0 \\ 0 & \psi_{22} & 0 \\ 0 & 0 & \psi_{33} \end{vmatrix}$$
2×2

Non-Recursive Models Approximated by A Dynamic Recursive System with Small Time Lag

Let us assume that inpatient service use (Y_1) and outpatient service use (Y_2) are affected by health or functional status (X) in the first quarter of a year. Multi-waves $(t_1, t_2, t_3 \ldots t_i)$ of quarterly utilization data will be gathered. A reciprocal relationship between the two utilization variables measured in the last wave of data collection is assumed (Figure 41).

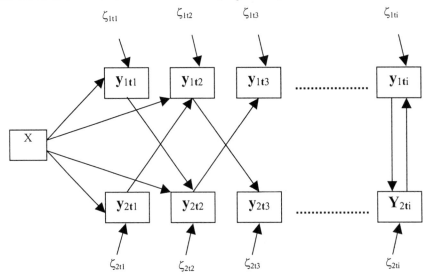

Figure 41. A Multi-Wave Panel Model with Two Utilization Variables

Structural equations are:

A generic model is: $\underset{\sim}{Y} = \underset{\sim}{B}\,\underset{\sim}{Y} + \underset{\sim}{\Gamma}\,\underset{\sim}{X} + \underset{\sim}{\zeta}$.

A model for a specific time point is: $Y_{ti} = \beta Y_{(ti-1)} + \Gamma X + \zeta_{ti}$.

For Y_1 at Time i: $\quad Y_{1\,ti} = \beta\, y_{2(ti-1)} + \Gamma X + \zeta_{1\,ti}$.

For Y_2: at Time i: $\quad Y_{2\,ti} = \beta\, y_{1(ti-1)} + \Gamma X + \zeta_{2\,ti}$.

AN APPLICATION OF PATH ANALYSIS WITH LISREL

The purpose of this example is to apply path analysis, using the LISREL VIII computer program, to examine how the factors of CASEMIX, %SURG, ALOS, BEDSIZE, and RNRATIO affect hospital mortality rate.

Data

Data for this example were based on the Health Care Financing Administration (HCFA) data, released in 1986, of the nation's hospitals having mortality rates significantly higher or lower than the national average. With the hospital as the unit of analysis, data were collected on the hospital organizational characteristics of CASEMIX, %SURG, ALOS, BEDSIZE, and RNRATIO to analyze their effects on hospital mortality rate (see Table 12).

Variables' Definition

Table 12. Operational Definition of the Variables

Variables	Label	Operational Definition
Y1	MORTR	Hospital mortality rate (per 1,000 Medicare patients per hospital)
X1	CASEMIX	HCFA's case mix index
X2	%SURG	Percent surgical patients
X3	ALOS	Average length of stay
X4	BEDSIZE	Hospital bed size
X5	RNRATIO	RNs per 100 nurses in a hospital

Analytic Method

This example uses the Linear Structural Relationships (LISREL) computer program to perform path analysis of the causal model depicted in Figures 42 and 43. The actual construction of the causal model precedes the statistical procedure and is based on the interpretation of current theory in the field. The researcher must state where causal relationships may exist between the variables and what the direction of that relationship may be. In essence, each linkage included represents implicitly a hypothesis that can be tested by estimating the magnitude of the relationship. A LISREL path analysis uses a structural equation model that specifies the causal relationships among a set of variables. Endogenous variables are those variables that are explained by the model, and that are specified as causally dependent on other endogenous variables and/or exogenous variables. Exogenous variables are those variables that are determined outside of the model.

The causal structure among the variables is expressed mathematically as follows: $Y = B_Y + \Gamma_X + \zeta$,

where:

Y is a vector of observed endogenous variables measured without errors;

X is a vector of observed exogenous variables measured without errors;

B is a matrix of coefficients relating the endogenous variables to one another;

128

Γ is a matrix of coefficients relating the exogenous variables to the endogenous variables; and

ζ is a vector of residual errors in equations, indicating that the endogenous variables are not perfectly predicted by the structural equations.

One of the main advantages of using LISREL for a path analysis is LISREL's hypothesis testing capability (Bollen, 1989). A X^2 likelihood ratio test is calculated for the null hypothesis that the sample covariance matrix S is drawn from a population characterized by the hypothesized covariance matrix Σ. If the null hypothesis is not rejected, the adequacy of the model specified by the researcher is confirmed.

Findings

The proposed structural model of hospital mortality is depicted by the diagram in Figure 42. The correlation matrix for the variables in this model can be found in Table 13.

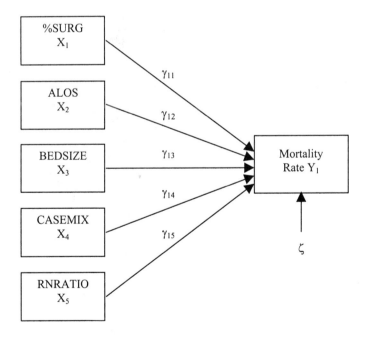

Figure 42. Proposed Structural Model

Table 13. Correlation Coefficients of the Study Variables

Variables	MORTR	CASEMIX	SURG	ALOS	BEDS	RNRA
MORTR	-.11					
CASEMIX	-.11	1.00				
SURG	-.25*	.21*	1.00			
ALOS	.38*	-.02	.21*	1.00		
BEDS	.12	.29*	.04	.20*	1.00	
RNRA	-.24*	.29*	.21*	.01	.20*	1.00

* Significant at .05 or lower level

Structural equations, the matrix form, and the LISREL program are as follows:

A Proposed Model:

$$Y_1 = \Gamma_{11}X_1 + \Gamma_{12}X_2 + \Gamma_{13}X_3 + \Gamma_{14}X_4 + \Gamma_{15}X_5 + \varsigma_1.$$

$$\left| Y_1 \right| = \left| \gamma_{11}\ \gamma_{21}\ \gamma_{31}\ \gamma_{41}\ \gamma_{51} \right| \ x\ \left| \begin{array}{c} X_1 \\ X_2 \\ X_3 \\ X_4 \\ X_5 \end{array} \right| + \left| \varsigma_1 \right|$$

LISREL Program:

```
Mortality and Specialization
DA NI = 6 NO = 244
LA
MORTR CASMX PER SUR ALOS BDSIZ RNRATIO
KM
```

130

```
1.00
-.11 1.00
-.25 0.21 1.00
.38 -0.02 .21 1.00
.12 0.29 0.04 0.20 1.00
-.24 0.29 0.21 0.01 0.2 1.00
SE
1 2 3 4 5 6
MO NY = 1 NX = 5 PS = DI
OU SE TV EF MI
```

A Revised Model:

$$Y_1 = B_{12} Y_2 + \Gamma_{11}X_1 + \Gamma_{12}X_2 + \Gamma_{13}X_3 + \Gamma_{14}X_4 + \varsigma_1$$
$$Y_1 = \Gamma_{21}X_1 + \Gamma_{22}X_2 + \Gamma_{23}X_3 + \Gamma_{24}X_4 + \varsigma_2$$

$$\begin{vmatrix} Y_1 \\ Y_2 \end{vmatrix} = \begin{vmatrix} 0 & B_{12} \\ 0 & 0 \end{vmatrix} \begin{vmatrix} Y_1 \\ Y_2 \end{vmatrix} + \begin{vmatrix} \gamma_{11} & \gamma_{12} & \gamma_{13} & \gamma_{14} \\ \gamma_{21} & \gamma_{22} & \gamma_{23} & \gamma_{24} \end{vmatrix} \times \begin{vmatrix} X_1 \\ X_2 \\ X_3 \\ X_4 \end{vmatrix} + \begin{vmatrix} \varsigma_1 \\ \varsigma_1 \end{vmatrix}$$

LISREL model: A revised model
```
Mortality and Specialization
DA NI = 6 NO = 244
LA
MORTR CASMX PER SUR ALOS BDSIZ RNRATIO
KM
1.00
-.11 1.00
-.25 0.21 1.00
.38 -0.02 .21 1.00
.12 0.29 0.04 0.20 1.00
-.24 0.29 0.21 0.01 0.2 1.00
SE
1 2 3 4 5 6
MO NY = 2 NX = 4 BE = SD PS = DI PH = SY,FR GA = FU,FR
FI GA(2,4)
OU SE TV EF
```

The squared multiple correlation coefficient for MORTR is .296, indicating that 29.6% of the variation can be explained by the model. The significant zeta value of .704 indicates that 70.4% of the variation in MORTR is explained by factors not included in the model. Table 14 gives a summary of statistics from the LISREL path analysis. It can be seen that there are significant gamma linkages for %SURG, ALOS, and RNRATIO, with ALOS being the strongest. However, there are small and insignificant linkages for BEDSIZE and particularly for CASEMIX. The decomposition of the causal components for the proposed model is also shown in Table 14. In this case, all causal effects are direct causal effects, as represented by the simple path between each of the exogenous variables and MORTR. In this model, there are no a priori fixed elements among the gammas. Therefore, the model is just identified and fits the data perfectly. However, the insignificant linkages for BEDSIZE and CASEMIX indicate that this simplistic model may need revision to better explain the variation in hospital mortality rate. Although the R^2 of 29.6% is moderately respectable, previous research on mortality rate suggests that the effects of the variables included in this model on mortality rate are more complex than are depicted in the proposed model. A revised model, shown in Figure 43, is therefore presented in an attempt to better apply the theory behind mortality rate and improve the previous model.

Table 14. The Decomposition of Effects: Proposed Structural Model

Endogenous Variables	Exogenous Variables	Total Association	Total Causal Effect	Direct Causal Effect	Indirect Causal Effect	Other Indirect Effect
Y	X1	r01=-.11	-.00675	-.00675		-.103
	X2	r02=-.25	-.30052	-.30052	-	.051
	X3	r03=.38	.42737	.42737	-	-.047
	X4	r04=.12	.08786	.08786	-	.032
	X5	r05=-.24	-.19677	-.19677	-	-.043

The direct linkages for ALOS, RNRATIO and %SURG were all maintained in the revised model. It is worth noting that different models were considered: all the models considered were theoretically grounded. In addition to having significant coefficients in the proposed model, some theoretical support can be found for these linkages. Chassin et al. (1989) state

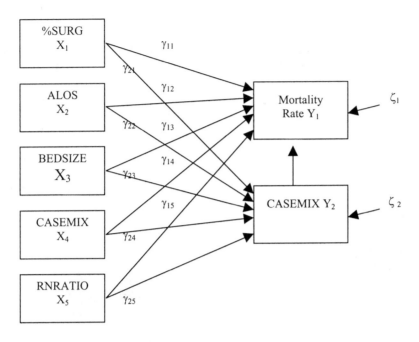

Figure 43. The Revised Model

that death rates in hospitals with longer ALOS tend to be higher. Farley and Ozminkowski (1992) found that nursing intensity (RNRATIO) to be negatively associated with mortality rate. They found mixed support for the hypothesis that the volume of certain procedures (%SURG) is negatively related to mortality rate. The relationship of BEDSIZE to mortality rate is less clear. Although many studies have found BEDSIZE to be related to mortality rate, this relationship has also been found to be dependent on other organizational variables (Al-Haider and Wan, 1991; Aiken, 1998; Aiken, Sloane, and Sochalski, 1998; Dubois et al., 1987). Moreover, the variable CASEMIX had the weakest causal linkage in the proposed model, indicating that the relationship between CASEMIX and mortality may also be complex. The squared multiple correlation for CASEMIX is 175, indicating that only 17.5% of the variance in this variable is explained by ALOS, BEDSIZE, RNRATIO, and %SURG. The significant zeta of .825 indicates that 82.5% of the variation is unexplained. The squared multiple correlation for MORTR stayed the same, indicating that 29.6% of the variation in MORTR is explained by the model, while 70.4% is explained by factors outside of the model.

133

Table 15 gives summary statistics for the path analysis of the revised model. The linkages from %SURG, ALOS, and RNRATIO to MORTR are still statistically significant, as are the linkages from BEDSIZE, %SURG, ALOS, and RNRATIO to CASEMIX. Interestingly enough, the linkage from BEDSIZE to MORTR is still insignificant in the revised model, while the linkage from BEDSIZE to CASEMIX is significant. The decomposition of causal effects for the revised model is also shown in Table 15. In the revised model there are no a priori fixed elements among the gammas, so the model is just identified. The coefficient of determination gives a moderate reliability. The X^2 likelihood ratio came out to be similar to that in the proposed model. Furthermore, a Q - plot of the normalized residuals shows that they all fall below the 45degree line. All of this indicates that although the fit of the model to the data is not too bad, the model could be further improved.

Table 15. The Decomposition of Effects: Revised Structural Model

Endogenous Variables	Exogenous Variables	Total Association	Total Causal Effect	Direct Causal Effect	Indirect Causal Effect	Other Indirect Effect
Y_1	X_2	r_{12}=-.25	-.302	-.302	0.00	.051
	X_3	r_{13}= .38	.428	.428	0.00	-.047
	X_4	r_{14}= .12	.086	.086	0.00	.032
	X_5	r_{15}=-.24	-.198	-.198	0.00	-.043
Y_2	X_2	r_{22}=-.21	.197	.181	.021	.027
	X_3	r_{23}=-.02	-.110	-.113	-.03	.09
	X_4	r_{24}= .29	.266	.265	-.001	.024
	X_5	r_{25}= .29	.2	-.198	.02	-.043

The proposed model is a generic model, showing only a direct path from each factor to MORT. The revised model attempted to improve the variation explained by creating a more intricate causal linkage from CASEMIX to MORTR. Generally, all model statistics for the proposed model stayed the same in the revised model, except for the improvement in the coefficient of determination and the appearance of a significant causal linkage from BEDSIZE to CASEMIX.

The literature suggests that there are three major categories of variables that affect the variation in hospital mortality rate. These are patient characteristics, hospital organizational characteristics, and community/contextual factors (Dubois et al., 1987; Wan, 1992). The factors included in the present example represent only the category of hospital organizational characteristics. In view of this, it seems reasonable to conclude that the R- square for the mortality rate model cannot be improved further without the addition of variables from the other two categories. The organizational variables could be improved by including other than only Medicare patients, by including the percentage of medical cases as well as of surgical cases, by extending the evaluation period to discharge plus 30 days, and by combining discharge data with disease-staging data.

Although the R^2 for the model cannot be improved without adding predictor variables, the fit of the revised model could easily be improved by examining more intricate relationships between the variables. Using LISREL, the example was able to overcome the limitation of traditional path analysis; in path analysis, goodness of fit testing can be done. To validate the model, the proposed model should be examined with a different sample universe.

CONCLUSION

The causal mechanisms responsible for the relationship between exogenous and endogenous variables should be explored under a theoretically specified model. Competing theories may have to be put forward to examine the determinants of outcomes and mechanisms of action that may influence patient care outcomes or organizational performance. LISREL modeling in path analysis can identify important factors that affect the variation in quality or outcomes. This analytical strategy should shed considerable light on the question of why hospitals with varying characteristics experience different mortality rates.

REFERENCES

Aiken, L.H. (1998). How organization and staffing of hospitals affect patient outcomes. *International Medical Forum* 103: 79-85.

Aiken, L.H., Sloane, D.M., Sochalski, J. (1998). Hospital organization and outcomes. *Quality of Health Care* 7(4): 22-226.

Al-Haider, A.S., Wan, T.T.H. (1991). Modeling organizational determinants of hospital mortality. *Health Services Research* 21: 303-323.

Bollen, K.A. (1989). *Structural Equations with Latent Variables.* New York: John Wiley & Sons.

Chassin, M.R., Park, R.E., Lohr, K.N., Keesey, J., Brook, R.H. (1989). Differences among hospitals in Medicare patient mortality. *Health Services Research* 24 (1): 1-31.

Dubois, R.W., Brook, R.H., Rogers, W.H. (1987). Adjusted hospital death rates: A potential screen for quality of Medicare care. *American Journal of Public Health* 77(9): 1162-1166.

Farley, D.E., Ozminkowski, R.J. (1992). Volume-outcome relationships and inhospital mortality: The effect of changes in volume over time. *Medical Care* 30: 77-94.

Wan, T.T.H. (1992). Hospital variations in adverse patient outcomes. *Quality and Utilization Review* 7: 50-53.

CHAPTER 8

STRUCTURAL EQUATION MODELS WITH LATENT VARIABLES

In this chapter, we consider building structural equation models with latent (unobservable) variables with measurement errors.

MODEL SPECIFICATION

In structural equation models, the latent variable model can be written as:

$$\eta = B\eta + \Gamma\xi + \zeta \qquad (8.1)$$

The component of the structural models is the measurement model which can be written as:

$$y = \Lambda y\, \eta + \varepsilon \qquad (8.2)$$
$$x = \Lambda x\, \xi + \delta \qquad (8.3)$$

Equations 8.2 and 8.3 are derived using confirmatory factor analysis, with some specifications such as $\Lambda y = Ip$, $\Lambda x = Iq$, $\Theta\delta = 0$, and $\Theta\varepsilon = 0$. Then the equation 8-1 can be rewritten (if no measurement models are involved) as:

$$Y = BY + \Gamma X + \zeta \qquad (8.4)$$

There are several different structural equation models with latent variables (Bollen, 1989; Byrne, B.M., 1998). For example, the Multiple Indicator Multiple Cause (MIMIC) model deals with the case of a single unobservable, theoretical variable (η_1), which is measured by multiple indicators and is influenced by multiple causes (Figure 44).

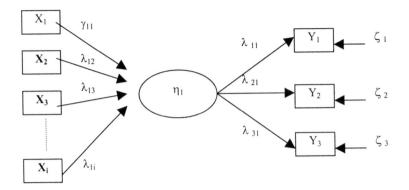

Figure 44. A MIMIC Model

Measurement model:

$$
\begin{vmatrix} Y_1 \\ Y_2 \\ Y_3 \end{vmatrix} = \begin{vmatrix} \lambda_{11} \\ \lambda_{21} \\ \lambda_{31} \end{vmatrix} \eta_1 + \begin{vmatrix} \varepsilon_1 \\ \varepsilon_2 \\ \varepsilon_3 \end{vmatrix}
$$

Structural equation model:

$$
\eta_1 = \begin{vmatrix} \gamma_1\ \gamma_2\ \gamma_3 \dots \gamma_i \end{vmatrix} x \begin{vmatrix} X_1 \\ X_2 \\ \cdot \\ \cdot \\ X_i \end{vmatrix} + \zeta
$$

$$
\eta = \gamma\, X + \zeta \qquad\qquad (8.5)
$$

Examples and specification:

A recursive model of health services use with social support, self-reported health and functional status factors as predictors (see Figure 45) and with attached program specifications. A nonrecursive model of health services use with social support, self-reported health and functional status factors as predictors can be examined if a reciprocal relationship between physician visits and use of social services is assumed.

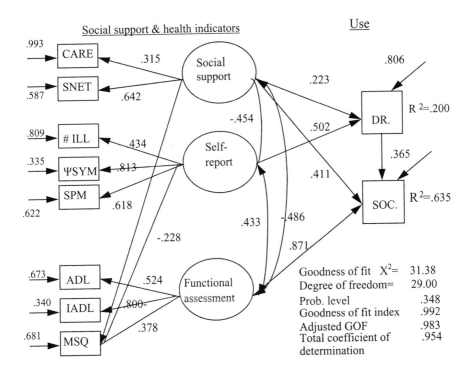

Figure 45. A Recursive Model of Health Services Use with Social Support Self Report Functional Status Factors as Predictors

Following is an example to illustrate how an exogenous latent variable (ξ) may influence an endogenous latent variable (η) as well as its indicator (Y_3). In addition, correlated error terms (δ_3 and ε_3) are assumed (Figure 46). Let us assume that functional outcomes as an endogenous latent variables with three indicators (physical, mental, and social functioning), and care interventions as an exogenous variable with three nursing care management activities (care planning, coordinated case management, and outcome tracking). This proposed model enables the investigator to explore the effect of care interventions on patient care outcomes. In addition, it allows asking a crucial question about the overall effects of nursing interventions on one specific functional outcome.

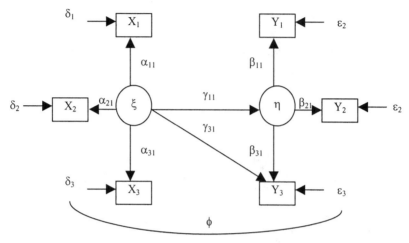

Figure 46. Model with Correlated Measurement Errors

LISREL specification of the model (Figure 46): Let all observed variables be y-variables and all latent variables be η-variables.

$$
\begin{vmatrix} X_1 \\ X_2 \\ X_3 \\ Y_1 \\ Y_2 \\ Y_3 \end{vmatrix} = \begin{vmatrix} 1 & 0 \\ \alpha_{21} & 0 \\ \alpha_{31} & 0 \\ 0 & 1 \\ 0 & \beta_{21} \\ 0 & \beta_{31} \end{vmatrix} \begin{vmatrix} \xi \\ \eta \end{vmatrix} + \begin{vmatrix} \delta_1 \\ \delta_2 \\ \delta_3 \\ \varepsilon_1 \\ \varepsilon_2 \\ \varepsilon_3 \end{vmatrix}
$$

$$
\underset{\sim}{Y} \quad = \quad \Lambda_y \qquad \underset{\sim}{\eta} \quad + \quad \underset{\sim}{\varepsilon}
$$

$$
\begin{vmatrix} \xi \\ \eta \end{vmatrix} = \begin{vmatrix} 0 & 0 \\ b & 0 \end{vmatrix} \begin{vmatrix} \xi \\ \eta \end{vmatrix} + \begin{vmatrix} \zeta_1 \\ \zeta_2 \end{vmatrix}
$$

$$
\underset{\sim}{\eta} \quad = \quad \underset{\sim}{B} \qquad \underset{\sim}{\eta} \quad + \quad \underset{\sim}{\zeta}
$$

$$\underset{\sim}{\Phi} = \begin{vmatrix} \sigma\xi^2 & 0 \\ 0 & \sigma\xi^2 \end{vmatrix}$$

$$\underset{\sim}{\Theta}_\varepsilon = \begin{vmatrix} \sigma\delta_1{}^2 & & & & & \\ 0 & \sigma\delta_2 & & & & \\ 0 & 0 & \sigma\delta_3{}^2 & & & \\ 0 & 0 & 0 & \sigma\varepsilon_1{}^2 & & \\ 0 & 0 & 0 & 0 & \sigma\varepsilon_2{}^2 & \\ 0 & 0 & 0 & 0 & 0 & \sigma\varepsilon_3{}^2 \end{vmatrix}$$

The LISREL Model

Measurement Model:

$$\underset{\sim}{Y} = \Lambda\underset{\sim}{\eta} + \underset{\sim}{\varepsilon}$$

pxm

$$\underset{\sim}{X} = \Lambda\underset{\sim}{\xi} + \underset{\sim}{\delta}$$

qxm

Structural Equation Model:

$$\underset{\sim}{\eta} = \underset{\sim}{\beta}\underset{\sim}{\eta} + \underset{\sim}{\Gamma}\underset{\sim}{\xi} + \underset{\sim}{\zeta}$$

mxm

Covariance Matrix

$\xi : \Phi\,(\underset{\sim}{n} \times \underset{\sim}{n})$ $\xi\,\Phi\,(\underset{\sim}{m} \times \underset{\sim}{m})$

$\varepsilon : \Theta\varepsilon\,(\underset{\sim}{p} \times \underset{\sim}{p})$ $\delta : \Theta\delta\,(\underset{\sim}{p} \times \underset{\sim}{p})$

An autoregression model is presented in Figure 47. This model enables to investigate the repeated measures of the same indicator (i.e., self-perceived health status) over four time points in a longitudinal study. The beta coefficient (ß) represents the relationship between the two measured indicators and the residual term (ζ) represents the unexplained variance for each indicator variable. The specification of this model is presented below.

141

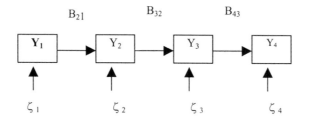

Figure 47. Autoregression
Note: There are no X variables in this model.

$$y_t = B_t\, y_{t-1} + \zeta_t. \qquad\qquad t = 2,3,4.$$

$$\underset{\sim}{B} = \quad \begin{array}{c} \\ \\ 4\times 4 \\ \\ \end{array} \begin{vmatrix} 0 & 0 & 0 & 0 \\ \beta_{21} & 0 & 0 & 0 \\ 0 & \beta_{32} & 0 & 0 \\ 0 & 0 & \beta_{43} & 0 \end{vmatrix}$$

$$\underset{\sim}{\Phi} = \mathrm{diag}\,(\,\sigma\xi 1^2,\,\sigma\xi 2^2,\,\sigma\xi 3^2,\,\sigma\xi 4^2\,).$$

The model is overidentified (df = 3).

A complex model with a reciprocal link between two latent endogenous variables is presented in Figure 48. For example, patient satisfaction with care (η_1) is assumed to be reciprocally related to nurses' job satisfaction (η_2). The clinical care integration mechanism (ξ_1) has a direct effect on both patient satisfaction and nurse satisfaction. Incentive payment (ξ_2) and clinical support system (ξ_3) are two exogenous latent variables that directly affect nurses' job satisfaction. The specification is presented below.

A measurement model for X:

$$\begin{vmatrix} X_1 \\ X_2 \\ X_3 \\ X_4 \\ X_5 \\ X_6 \\ X_7 \end{vmatrix} = \begin{vmatrix} 1 & 0 & 0 \\ \lambda_2 & 0 & 0 \\ \lambda_3 & \lambda_4 & 0 \\ 0 & 1 & 0 \\ 0 & \lambda_6 & 0 \\ 0 & 0 & 1 \\ 0 & 0 & \lambda_7 \end{vmatrix} X \begin{vmatrix} \xi_1 \\ \xi_2 \\ \xi_3 \end{vmatrix} + \begin{vmatrix} \delta_1 \\ \delta_2 \\ \delta_3 \\ \delta_4 \\ \delta_5 \\ \delta_6 \\ \delta_7 \end{vmatrix}$$

142

A measurement model for Y:

$$
\begin{vmatrix} Y_1 \\ Y_2 \\ Y_3 \\ Y_4 \end{vmatrix} = \begin{vmatrix} 1 & 0 \\ \lambda_1 & 0 \\ 0 & 1 \\ 0 & \lambda_2 \end{vmatrix} X \begin{vmatrix} \eta_1 \\ \eta_2 \end{vmatrix} + \begin{vmatrix} \varepsilon_1 \\ \varepsilon_2 \\ \varepsilon_3 \\ \varepsilon_4 \end{vmatrix}
$$

A structural equation model:

$$
\begin{vmatrix} \eta_1 \\ \eta_2 \end{vmatrix} = \begin{vmatrix} 0 & \beta_{12} \\ \beta_{21} & 0 \end{vmatrix} X \begin{vmatrix} \eta_1 \\ \eta_2 \end{vmatrix} + \begin{vmatrix} \gamma_{11} & \gamma_{12} & 0 \\ \gamma_{21} & 0 & \gamma_{23} \end{vmatrix} X \begin{vmatrix} \xi_1 \\ \xi_2 \\ \xi_3 \end{vmatrix} + \begin{vmatrix} \zeta_1 \\ \zeta_2 \end{vmatrix}
$$

Other parameter matrix

$$
\underset{\sim}{\Phi} = \begin{vmatrix} \phi_{11} & & \\ \phi_{21} & \phi_{22} & \\ \phi_{31} & \phi_{32} & \phi_{33} \end{vmatrix}
$$

$$
\underset{\sim}{\psi} = \begin{vmatrix} \psi_{11} & \\ \psi_{21} & \psi_{22} \end{vmatrix}
$$

$$
\underset{\sim}{\Theta}_\varepsilon = \text{diag} (\sigma\varepsilon_1{}^2, \sigma\varepsilon_2{}^2, \sigma\varepsilon_3{}^2, \sigma\varepsilon_4{}^2).
$$

$$
\underset{\sim}{\Theta}_\delta = \text{diag} (\sigma\delta_1{}^2, \sigma\delta_2{}^2, \sigma\delta_3{}^2, \sigma\delta_4{}^2).
$$

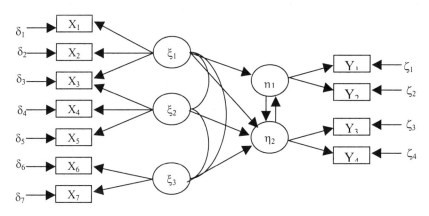

Figure 48. A Nonrecursive Model with Latent Constructs for Exogenous and Endogenous Variables

This model can be illustrated by an example to assess the effects of three domains of health informtics (connectivity, applicability, and communicability) on hospital quality (η_1) and efficiency (η_2). The relationship between quality and efficiency is assumed to reciprocal in this diagram. The three domains of health informatics are assumed to be correlated and have different effects on each of the latent endogenous variables.

APPLICATIONS

Problem

Hospital performance is very difficult to measure. Hospitals are complex organizations, performing widely varied tasks and services, and employing professionals with diverse training. Griffith (1987) classifies the measurement of hospital performance into three broad categories: 1) demand measures such as counts, time distribution, and market share; 2) outcome measures such as outputs, efficiency and quality; and 3) resource measures such as costs, revenues, and profits. For the purpose of this example, performance can be viewed as a function of efficiency and effectiveness (Wan and Shukla, 1987; Wan, Clement, and Spotswood, 1995). The term *efficiency* can be explained in three concepts: (1) economic efficiency measured in terms of resources consumed per unit of services or output; (2) response efficiency, which refers to how quickly a system responds or reacts to a request for service; (3) production efficiency, which refers to the production system's outputs per unit of time (Chern and Wan, 2000; Hollingsworth, Dawson, and Maniadakis, 1999; Wang, et al., 1999;Youn and Wan, 2001).

Effectiveness refers to the quality of output, as well as and the financial viability (Wan, 1995; Lin and Wan, 1999). In industry, quality can be measured by product failure rate and consumer satisfaction. In health care, quality cannot be defined easily. However, certain conspicuous aspects of the quality of health care certainly can be defined and measured: post-surgical infection rates, repeated hospitalizations, and mortality rates.

The relationship between different indicators of hospital performance is important. The magnitude and the direction of such a relationship can contribute significant findings to the literature on hospital performance and thus enhance our understanding. This application examines hospital performance, using LISREL. Several performance indicators will be used to structure a LISREL model proposed to explain the variation in hospital performance (Table 16).

Methodology

Data for this application were obtained from three data sources:
1. The American Hospital Association (AHA)
2. Virginia Hospital Cost Review Commission
3. The Medical Society of Virginia Review Organization

Table 16. List of the Study Variables and Definitions

	Variable	Definition
	Exogenous variables	
X1	BEDSIZE	Hospital bed size
X2	HITECH	Number of critical care specialty services offered, including open heart surgery, organ transplant, etc.
X3	CASEMIX	HCFA DRG - based hospital case mix index
X4	SEVERITY	% patients treated in the special care units
X5	METRO SIZE	Metropolitan area size
X6	MULTI	Multi-hospital system coded 1; else =0
X7	MDSCL	Medical school affiliation coded 1; else=0
X8	FORPROFIT	For-profit hospital coded 1; else=0
	Endogenous variables	
Y1	COST	Cost efficiency: Average charges per patient discharged
Y2	TECHEFF	Technical efficiency: DEA measure
Y3	ALOS	Product efficiency: measured by average length of stay
Y4	%SHARE (η_2)	% Medicare patients served by a hospital in an area
Y5	NETPROFIT (η_3)	Amount of net income earned by a hospital
Y6	TRAUMAR*	In-hospital trauma rate
Y7	MEDPROBR*	Rate of discharges with unstable medical conditions
Y8	TXPROBR*	Rate of treatment problems
Y9	COMPRATE*	Complication rate
Y10	DEDPROR*	Rate of unexpected deaths
Y11	NTRAUMAR	Repeated measure of TRAUMAR
Y12	NMEDPROBR	Repeated measure of MEDPROBR
Y13	NTXPROBR	Repeated measure of TXPROBR
Y14	NCOMPRATE	Repeated measure of COMPRATE
Y15	NDEDPROR	Repeated measure of DEDPROR

Notes: * Measured in 1987.
Repeated measures were done in 1988.
Inefficiency is a measure of deficiency

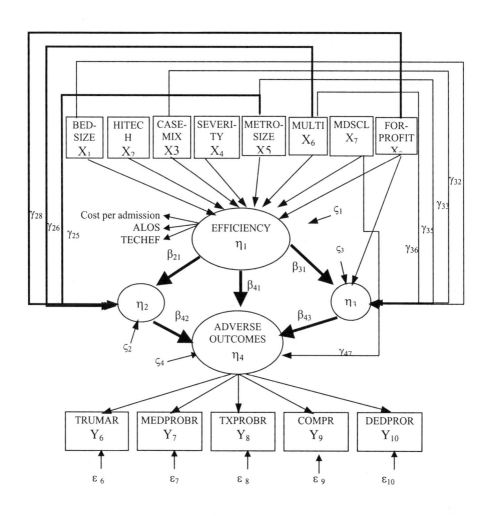

Figure 49. A Proposed Structural Equation Model of Hospital Performance

A LISREL program for the measurement model of adverse patient outcomes is presented below:

```
CAUSAL MODEL OF ORGANIZATIONAL DETERMINANTS OF
PERFORMANCE
DA NI=26 NO=85 MA=KM
LA
'BEDSIZE' 'EXPEND' 'OCC' 'HITECH' 'BRDCERT' 'PCTSURG'
'PCTRN' 'CASEMIX'
'COST' 'ALOS' 'SEVERity' 'METROSIZ' 'TRAUMAR' 'MEDPROBR'
```

146

'TXPROBR' 'COMPRATE' 'DEDPROBR' 'NETPROR' 'INEFF87'
'TECHEFF' '%SHARE' 'FTEBED' 'READRATE' 'MULTISYS'
'MEDSCHOO' 'OWNER'

KM SY

```
 1.00
  .90  1.00
  .55   .66  1.00
  .86   .85   .54  1.00
  .30   .40   .37   .32  1.00
  .72   .82   .53   .72   .54  1.00
  .29   .43   .22   .30   .27   .37  1.00
  .81   .76   .53   .73   .36   .79   .39  1.00
  .56   .52   .25   .47   .14   .50   .43   .57  1.00
  .51   .64   .47   .54   .22   .59   .25   .51   .41
 1.00
  .35   .36   .19   .31  -.06   .25   .16   .21   .30
  .29  1.00
  .36   .58   .39   .42   .16   .55   .57   .47   .47
  .57   .16  1.00
 -.044 -.046 -.02  -.008  .05  -.037 -.14  -.037 -.175
  .06  -.065 -.18  1.00
 -.203 -.181 -.197 -.169 -.29  -.239 -.172 -.26  -.158
  .007  .026 -.122  .40
 1.00
 -.095 -.078 -.12  -.07  -.27  -.09   .05  -.04   .057
  .24   .066  .17   .31
  .34  1.00
 -.06  -.079 -.084 -.04  -.34  -.06  -.01   .013  .012
  .14   .062  .14   .20
  .30   .76  1.00
 -.08  -.08  -.09   .01  -.28  -.001  .015 -.016  .055
  .25   .007  .216  .217
  .362  .61   .59  1.00
  .43   .43   .33   .42   .294  .45   .15   .41   .11
  .25   .19   .03   .068
 -.14  -.18  -.13  -.13  1.00
  .29   .18   .01   .25  -.13   .22  -.03   .33   .36
  .27   .11   .16   .24
  .14   .37   .38   .44  -.03  1.00
  .18   .10  -.04   .24  -.09   .19   .01   .20   .35
  .18   .17   .11   .19
  .06   .20   .30   .27   .14   .58  1.00
 -.24  -.40  -.15  -.25  -.14  -.43  -.32  -.34  -.34
```

```
-.39   -.07   -.62   -.01
-.04   -.20   -.16   -.26    .05   -.22   -.20   1.00
 .43    .49    .45    .46    .28    .45    .29    .50    .30
 .28    .30    .28    .10
-.16   -.08   -.07   -.02    .13    .38    .22   -.12   1.00
 .005  -.12   -.25   -.08   -.42   -.32   -.09   -.07    .07
-.31    .15   -.29    .01
 .33    .14    .12   -.05   -.11    .13    .11    .09   -.03
1.00
 .14    .14    .08    .12    .098   .11    .24    .17    .18
 .07   -.08    .14   -.02
 .08   -.04   -.08   -.10   -.045   .01   -.10   -.28   -.14
 .06   1.00
 .72    .67    .44    .63    .24    .55    .29    .63    .40
 .38    .27    .38    .02
-.22   -.08   -.04   -.09    .29    .22    .13   -.30    .41
-.07    .21   1.00
-.03    .02   -.11   -.05   -.077  -.04    .17    .006   .015
 .017  -.09    .035  -.13
 .21   -.12   -.02   -.03    .13   -.06    .001  -.165  -.21
 .24    .48   -.13   1.0
SD
166.3 0.91 12.75 3.14 18.0 8.63 14.78 0.12 885.64 1.54
3.89 2.4 10.48 7.18 1.41
6.23 1.79 368.7 0.14 0.12 0.40 0.77 19.28 0.50 0.41 0.35
SE
14 15 16 17/
MO NX=4 NK=1 PH=sy,fr
LK
'ADVRSOUT'
FR LX(1,1) LX(3,1) LX(4,1)
ST 1 LX(2,1)
OU SE TV  MI SS RS eff   AD=OFF
```

Findings

The proposed covariance structural model of hospital performance is depicted by the diagram in Figure 49. The correlation between the study variables is shown in the above LISREL program. Before analysis of the covariance structure, the two separate measurement models were analyzed. The first measurement model was constructed by linking COST, TEFF, and ALOS variables to a latent construct called INEFFICIENCY, or lack of efficiency.

The variable ALOS was proposed to have the largest effect on this latent construct. Thus, this variable was set to one in the LISREL measurement program. Table 17 shows lambda's values. The total coefficient of determination for this measurement model was .814, indicating that the model explains 81.4 % of the variation in the latent construct. This model is a just identified model.

Table 17. A Measurement Model of Hospital Performance Indicators

Parameter	Lambda Y	Indicator	Construct
1,1	.951*	Cost efficiency (Cost)	Efficiency
2,1	.389*	Technical efficiency (EFF)	Efficiency
3,1	1.00*	Product efficiency (ALOS)	Efficiency
4,2	1.00*	Market share (%SHARE)	Market share
5,3	1.00*	Net profit (NTPRFT)	Financial viability
6,4	.437*	Rate of discharge w/unstable condition (MEDPROBR)	Adverse outcomes
7,4	1.00*	Treatment problem rate (TXPROBR)	Adverse outcomes
8,4	.938*	Postoperative complication rate (COMPRATE)	Adverse outcomes
9,4	.784*	Unexpected death rate (DEDPROBR)	Adverse outcomes

*Significance at .05 or lower level

A second measurement was constructed by linking TRAUMAR, MEDPROBR, TXPROBR, COMPR, and DEDPROBR to the latent construct ADVERSE OUTCOMES. The TXPROBR was set to one in the LISREL measurement program. A revised measurement model was constructed to improve the fit of the data. This was done by eliminating the variable TRAUMAR, which had insignificant loading on the common factor, adverse patient outcomes. Trauma rate may not be directly related to adverse outcomes, but may correlate with other variables. Although the correlation matrix does not show that trauma rate is significantly correlated with the other endogenous variables in the model, the study presumes that this variable is highly correlated with other variables in the model. Table 18 shows lambda's values for this measurement model. The X^2 likelihood ratio $(X^2/d.f)$ for this model was 1.02, with p =. 360, and the adjusted goodness of fit was .94, indicating that the model is excellent. After consideration of different covariance structural models that are theoretically grounded, the final revised model was constructed.

Factor and Structural Equations for the Model of Hospital Performance

Factor Equations - Measurement Model for Efficiency:

$$Y_1 = \lambda_{y11}\, \eta_1 + \varepsilon_1.$$
$$Y_2 = \lambda_{y21}\, \eta_1 + \varepsilon_2.$$
$$Y_3 = \lambda_{y31}\, \eta_1 + \varepsilon_3.$$

Factor Equations - Measurement Model for Adverse Outcomes:

$$Y6 = \lambda y64\, \eta4 + \varepsilon6.$$
$$Y7 = \lambda y74\, \eta4 + \varepsilon7.$$
$$Y8 = \lambda y84\, \eta4 + \varepsilon8.$$
$$Y9 = \lambda y94\, \eta4 + \varepsilon9.$$
$$Y10 = \lambda y104\, \eta4 + \varepsilon10.$$

Structural Equations (Figure 50): ξ is actually X in this model

$$\eta_1 = \gamma_{11}\, \xi_1 + \gamma_{12}\, \xi_2 + \gamma_{31}\, \xi_3 + \gamma_{14}\, \xi_4 + \gamma_{15}\, \xi_5 + \gamma_{16}\, \xi_6 + \zeta_1.$$
$$\eta_2 = B_{21}\, \eta_1 + \gamma_{25}\, \xi_5 + \zeta_2.$$
$$\eta_3 = B_{31}\, \eta_1 + B_{32}\, \eta_2 + \gamma_{32}\, \xi_2 + \gamma_{33}\, \xi_3 + \gamma_{35}\, \xi_5 + \zeta_3.$$
$$\eta_4 = B_{41}\, \eta_1 + B_{42}\, \eta_2 + B_{43}\, \eta_3 + \gamma_{46}\, \xi_6 + \zeta_4.$$

The revised model was constructed by eliminating two of the exogenous variables, which had insignificant loadings on the latent endogenous variables. Summary statistics for the covariance structural model can be found in Table 18. All statistical results indicate that the fit of this model was reasonably good. However, the adjusted goodness of fit index indicates that the model needs to be improved further.

Discussion

The final revised model is depicted in Figure 50. The revised model eliminated two exogenous variables that had insignificant loadings on the latent endogenous variables. From the business point of view, the literature finds no difference between for-profit and not-for-profit hospitals; therefore the for-profit (FRPRFT) variable was dropped from the model. In addition, joining a multi-hospital system may not affect hospital efficiency, because system affiliation does not directly interfere with a hospital's internal operations. Consequently, the variable that measured multi-hospital system membership was dropped from the model. Although the variable that measured teaching status had insignificant linkage with the efficiency construct, it was retained in the model because of its effect on adverse patient outcomes.

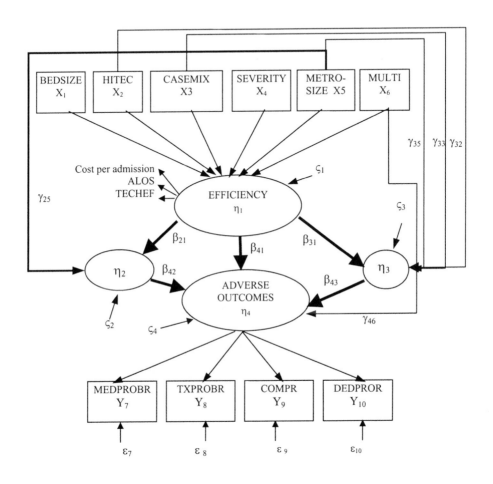

Figure 50. Revised Structural Equation Model of Hospital Performance

As shown in Table 18, the GOF index for this model was .910, and the likelihood X^2 ratio was 1.06, indicating that the model fits the data well. However, the adjusted GOF index indicates that the model needs improvement. The total coefficient of determination for the structural model was .921, indicating that 92 % of the variation in hospital performance can be explained by the current model. All gamma linkages were significant except for the linkages between case mix and high-tech, and efficiency.

Table 18. Structural Equation Model of Hospital Performance: Parameter Estimates

Endogenous Variable	Exogenous Variables					
	BED SIZE	HITECH	CASE MIX	SEVERITY	METRO SIZE	MDSCHL
Efficiency	.163*	.062	.187	.144*	.325*	--
Market share	--	--	--	--	-.575*	--
Financial viability	--	.257*	.288*	--	-.27*	--
Adverse outcomes	--	--	--	--	--	-.171*

-- The variable is not included in the analysis.
* Significance at .05 or lower level

Goodness of Fit (GOF) Statistics

	X^2 (with 63 degrees of freedom)	69.5
	GOF index	.91
	Adjusted GOF index	.835
	Root mean square residual	.07

The LISREL output indicated that the ξ matrix is not positive definite; however, this may be due to a normality problem. Metropolitan size showed a negative causal linkage with both market share and financial viability, indicating that hospitals in large metropolitan areas may be less financially strong and have lower market shares than is the cases for hospitals in smaller metropolitan areas. Hospitals in large metropolitan areas are likely to have many competitors, which would affect their market shares.

CONCLUSION

After two measurement models were run, the revised model was made by eliminating two of the eight exogenous variables proposed to affect efficiency. Although the likelihood X^2 ratio of the revised model was 1.06, and the total coefficient of determination of the structural model was .921, the current structural model needs further improvement. However, data normality problems may have caused some estimation problems and restricted further improvements.

The current model successfully demonstrates that concepts like hospital performance can be studied by using multiple indicators. But a few limitations should be noted. The availability of adverse patient outcome data for only a selected group of hospitals may restrict the study's comprehensive examination of the variation in hospital performance. Another limitation is that the stability of the structural model over time has not been tested. That would require panel data. Finally, because process indicators of quality for hospitals are not readily available, we are unable to explore the important relationship between efficiency and quality of care. Future work should incorporate patient, organizational, and technological indicators in a comprehensive model of hospital performance. In addition, the future study should seek a larger sample size, to enhance the capability of the modeling test to tease out potential confounding factors that may influence hospital performance.

REFERENCES

Bollen, K.A. (1989). *Structural Equations with Latent Variables.* New York: John Wiley & Sons.

Byrne, B.M. (1999). *Structural Equation Modeling with LISREL, PRELIS, and SIMPLIS: Basic Concepts, Applications, and Programming.* New Jersey: Lawrence Erlbaum Associates, Publishers.

Chern, J.Y., Wan, T.T.H. (2000). The impact of the prospective payment system on the technical efficiency of hospitals. *Journal of Medical Systems* 24(3): 159-172.

Griffith, J.R. (1987). *The Well-Managed Community Hospital.* Ann Arbor, Michigan: Health Administration Press.

Hollingsworth, B., Dawson, P.J., Maniadakis, N. (1999). Measurement of health care: a review of non-parametric methods and applications. *Health Care Management Research* 2(3): 161-172.

Lin, B.Y.J., Wan, T.T.H. (1999). Analysis of integrated health care networks' performance: A contingency-strategic management perspective. *Journal of Medical Systems* 23(6): 467-485.

Wan, T.T.H., Shukla, R. (1987). The contextual and organizational correlates of the quality of nursing care in community hospitals. *Quality Review Bulletin* 13(2): 61-64.

Wan, T.T.H., Clement, D.G., Spotswood, M. (1995). Assessing effectiveness and efficiency of hospital performance. In T.T.H. Wan, Analysis *and Evaluation of Health Care Systems: An Integrated Approach to Managerial Decision Making.* Baltimore: Health Professions Press.

Wang, B.B.L., Ozcan, Y.A., Wan, T.T.H., Harrison, J. (1999). Trends in hospital efficiency among metropolitan markets. *Journal of Medical Systems* 23(2): 83-97.

Youn, K.I., Wan, T.T.H. (2001). Effects of environmental threats on the quality of care in acute care hospitals. *Journal of Medical Systems* 25(5): 319-331.

CHAPTER 9

COVARIANCE STRUCTURE MODELS

The covariance structure model overcomes certain weaknesses of both factor analysis and structural equation models by merging them into a single model that simultaneously estimates latent variables from observed variables and the structural relations among the latent variables.

MATHEMATICAL MODEL OF A COVARIANCE STRUCTURE MODEL

Figure 51 illustrates the causal relationship of two exogenous latent variables to two endogenous latent variables. The two exogenous latent variables are not correlated. The two residual terms (ζ_1 and (ζ_2) are not correlated. The following equation is constructed:

$$\eta = \beta\eta + r\,\xi + \zeta \qquad (9.1),$$

where:

η is a (r x 1) vector of latent, endogenous variable;

ξ is a (s x 1) vector of latent exogenous variables;

ζ is a (r x 1) vector of errors in equations;

β is a (r x r) matrix of coefficients relating the endogenous variables to one another; and

Υ is a (r x s) matrix of coefficients relating the exogenous variables to the endogenous variables.

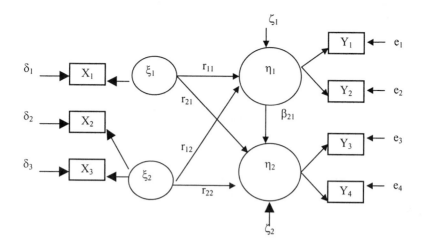

Figure 51. A Covariance Structure Model

A SEQ Model with correlated measurement errors and residuals can be assumed and tested in covariance structure analysis. A complex model, presented in Figure 52, that specifies a reciprocal relationship between two endogenous latent variables with correlated residual terms was formulated by Jöreskog and Sörbom (1993). They specify the following: There is a reciprocal (causal) relationship between two endogenous variables ($\eta 1$ and $\eta 2$). Two residual terms ($\zeta 1$ and $\zeta 2$) are correlated. This implies that an unidentified endogenous factor may share the common variance between the two structural equations. Correlated measurement errors of the endogenous variable are indicated.

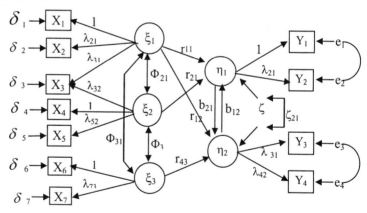

Figure 52. A Complex Model with Correlated Residuals

THREE EXAMPLES OF COVARIANCE STRUCTURE MODEL

Example 1: Causal Model of Physician Productivity

The first example illustrates how a community hospital's physician productivity variables are affected by the physicians' commitment to admitting their patients to that hospital and by their concerns ABOUT the hospital's support services, when control variables are simultaneously considered in a covariance structure equation model (Figure 53). The structural relationships among four latent variables are postulated.

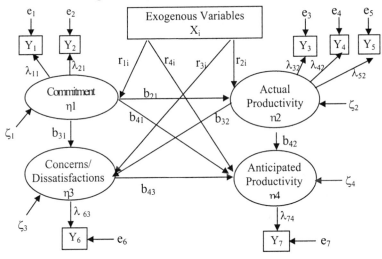

Figure 53. A Proposed Causal Model of Physician Productivity in a Community Hospital

Y_1 = Primary Facility for Admission.
Y_2 = Attending Status.
Y_3 = Average Monthly Patient Admissions.
Y_5 = Monthly Net Income in Inpatient Services.
Y_6 = Composite Index of Dissatisfaction.
Y_7 = Potential Patient Admissions.

Notes: Variables include age, hospital privileges, teaching status, surgery, location, and hours worked.

157

Table 19 summarizes the results of the covariance structure model. Physicians' commitment to the hospital is positively influenced by physician age and the office proximity to the hospital. Actual productivity of physicians (e.g., number of patients admitted by a physician to that community hospital) is positively affected by their commitment to admitting their patients to that hospital and by their actual productivity in a previous year, but is negatively related to their teaching status (physicians affiliated with a teaching medical center in the locality). The anticipated productivity of physicians is positively associated with their office location and their commitment to the hospital. Furthermore, more than 46% of the variance in the two productivity measures is accounted for by their respective predictor variables in the equations, 0.456 for the actual productivity and 0.789 for the anticipated productivity. Dissatisfaction with hospital support services, as measured by the number of concerns reported by physicians, exerts no relationship to other latent variables. The model fit statistics indicate that this model is reasonably fitted.

Table 19. Standardized Parameter Estimates for the Causal Model of Physician Productivity in a Community Hospital

Endogenous Variables	Predetermined Variables								
	X1 Age	X2 Hospital Privileges (Number)	X3 Teaching	X4 Surgeon	X5 Location	n1	n2	n3	R^2
Commitment to the Hospital (n1)	0.240*	-0.160	0.00	0.00	0.276*	-	-	-	0.135
Actual Productivity (n2)	0.151	-0.100	-0.283*	0.00	0.173	0.628*	-	-	0.456
Dissatis-faction or Concerns (n3)	0.024	-0.016	-0.437*	-0.259	0.487*	0.202	-0.163	-	0.586
Anticipated Productivity (n4)	0.086	-0.057	-0.241	-0.024	0.244*	0.257	0.997*	0.093	0.789

* Indicates $p < 0.05$. $X^2 = 70.08$, d.f. = 41; Goodness of Fit (GOF) Index = 0.901; Adjusted GOF = 0.811.

158

Example 2: Analytical Model for Investigating Oral-Facial Pain

The second example illustrates exogenous factors hypothesized to directly affect stress, oral-facial pain, and illness behavior. Figure 54 presents a theoretical model of the relationship between life stress and illness, using a relatively well-studied social stress theory (Wan, 1982; Lin, 1990). Social support and other exogenous variables exert direct effects on life stress, oral-facial pain, and illness behavior.

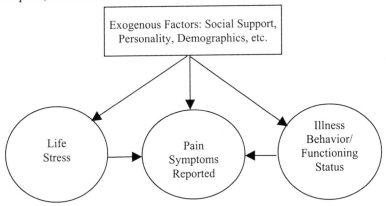

Figure 54. An Epidemiological Model of the Oral-Facial Pain Syndrome

Several testable hypotheses are formulated as follows:
1. Hypothesis 1: There is a direct causal effect of life stress on the degree of pain symptoms reported by the study patients.
2. Hypothesis 2: There is a direct causal effect of pain symptoms on the use of dental services and on the degree of functional incapacity of the study patients.
3. Hypothesis 3: Social support is a mediating factor between life stress and oral-facial pain, and between oral-facial pain and illness behavior (use of dental services or functional incapacity) of the study patients.

The instruments to measure specific variables or constructs are presented in Table 20. A measurement model of pain symptoms has been formulated and evaluated. The confirmatory factor analysis reveals that pain

159

associated with the dental problem studied has three related domains: thermo-pain, affect-based pain, and sensory-related pain (Figure 55). Because these three dimensions of oral-facial pain could be conceived as part of the second-order factor, a second-order factor of oral-facial pain symptoms could be postulated and evaluated. However, this proposed model did not fit the data well. Thus, only two related constructs of the oral-facial pain syndrome were included and tested (Figure 56). The sensory and affective aspects of the pain syndrome were positively associated (0.85). This model with two factorial dimensions of the oral-facial syndrome is well fitted. The structural relationships among stress, pain, and illness behavior can be further validated, using covariance structure modeling (Figure 57).

Table 20. Operational Definition & Measurement Instruments for the Study Variables

Study Variable	Operational Definition	Measurement or Instrument
Stress	Life events that are experienced by the respondent may be considered as stress-inducing factors, irrespective of the desirability of changes in the life span.	A 138-item inventory of life events questionnaire (Chiriboga, 1982)
Social Support	Support that is provided by spouse, children, siblings, parents, friends, neighbors, or other individuals is either material (instrumental) or non-material (expressive/emotional).	Social support network scale (Wan, 1985); Social Support Scale (Lin & Dean, 1984)
Health Function Status	Physical and mental functioning levels, the major domains of health status, can be indicated by the presence or absence of health disorders or symptoms in an adult population.	SF-36 (Ware et al., 1992); General Well-Being Index (Wan & Livieratos, 1978); NIMH Depression Scale (Radloff, 1977); Portable Psychiatric Symptom Scale (Pfeiffer, 1975)
Pain Symptom	MPD syndrome can be quantitatively assessed by using a psychophysiological measure developed in a laboratory experiment (Harkins, et al., 1986).	Pain symptom and severity scale (Harkins et al., 1986); Hopkins symptom checklist (Derogatis & Cleary, 1977)

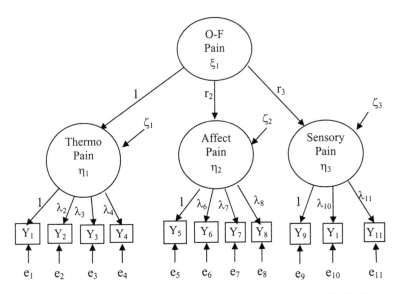

Figure 55. A Proposed Measurement Model of Oral-Facial Pain: A Second Order Factor

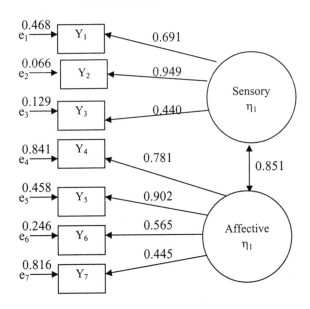

Figure 56. A Measurement Model of the Oral-Facial Pain Syndrome

Model Fitting Statistics:
   Chi-Square = 11.62 (p = 0.211), 12 d.f.
   Goodness of Fit Index = 0.943.
   Adjusted GFI = 0.868.
Notes:
   Y1 = Lowest Intensity Pain Measure (Sensory).
   Y2 = Usual Intensity Pain Measure (Sensory).
   Y3 = Highest Intensity Pain Measure (Sensory).
   Y4 = Lowest Intensity Pain Measure (Bothersome).
   Y5 = Usual Intensity Pain Measure (Bothersome).
   Y6 = Highest Intensity Pain Measure (Bothersome).

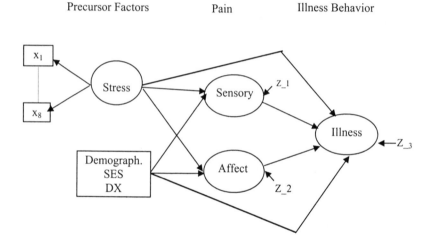

Figure 57. Structural Equation Model of the Oral-Facial Syndrome

Example 3: A Covariance Structure Model with an Interaction Effect

Statistical interactions can be postulated in accordance with the theoretical specifications of the study variables. It is a well-known fact that caregivers' burden can be affected by their stress coping mechanisms and their levels of distress associated with caregiving (Figure 58). The measurement model of the two exogenous latent constructs is evaluated and indicates that only four statistically significant indicators are valid for this model (Table 21). Two indicators (BEHAV and EFF) are excluded from the further development of a covariance structure model. The measurement model of caregiving burden is also evaluated, and the results are presented in Table 22. The independent effects on caregiving burden of the coping mechanisms and the distress levels of the caregivers are presented in Table 23. Caregivers' burden is negatively influenced by coping mechanisms (-0.286) and positively influenced by distress levels. Both latent variables account for 42% of the variance in caregiving burden, with a residual coefficient of 0.58.

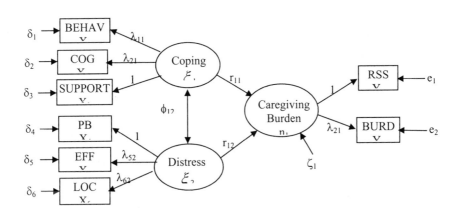

Figure 58. A Covariance Structure Model of Caregiving Burden

Y_1 (RSS) = Relatives' Stress Scale (Green, Smith, Gardiner, & Timbury, 1982).
Y_2 (BURDEN) = Zarit Burden Interview (Zarit, Reever, & Bach-Peterson, 1980).
X_1 (BEHAV) = Behavioral Coping Checklist.
X_2 (COG) = Cognitive Coping Checklist.
X_3 (SUPPORT) = Social Support Interview.
X_4 (PB) = Memory and Behavior Problems Checklist (Zarit and Zarit, 1982).
X_5 (EFF) = Perceived Self-Efficacy.
X_6 (LOC) = Perceived Loss of Control.
X_7 (BUFFER) = A statistical interaction term between X_3 and X_4.

According to the stress theory (Lin, 1992), stress and coping mechanisms can have a buffering effect on caregiving burden. This buffering hypothesis can be examined by introducing an interaction term, using the cross-product term (PB * SUPPORT) of two stronger indicators of the two latent constructs (coping and distress). The results are presented in Table 24 for the measurement model with an interaction term or buffering factor, and in Table 25 for the covariance structure model with a buffering factor. The distress factor is the only statistically significant variable to account for the variance in caregiving burden. However, the effect of the buffering factor is not statistically significant when the main effects of distress and coping mechanisms are controlled. The comparative model fit statistics of the main effect model and the buffering effect model can be found in Table 26. The analysis indicates that the buffering effect of the two indicators of distress and coping mechanisms is not supported by the data.

Table 21. A Measurement Model of Two Exogenous Latent Variables (Factor Loadings)

Indicators		Coping	Distress
BEHAV	X_1	0.170	
COG	X_2	0.618*	
SUPPORT	X_3	0.653*	
PB	X_4		0.817*
EFF	X_5		-0.189
LOC	X_6		0.570*

* $p < 0.05$ (Standardized regression coefficients)

Table 22. A Measurement Model of Endogenous Variables (Factor Loadings)

Indicators	Caregiving Burden
RSS y_1	0.998*
BURD y_2	0.601*

* $p < 0.05$

Table 23. A Structural Equation Model of the Effects of Exogenous Variables on Caregivers' Burden ($\eta 1$)

Construct		Standardized Regression Coefficient
Coping	γ_{11}	-0.286*
Distress	γ_{12}	0.591*
Errors in Equation ($\zeta 1$)		0.580*

*$p < 0.05$

Table 24. A Measurement Model of Exogenous Latent Variables (with a Buffering Effect)

Indicators		Coping	Distress	Buffer
BEHAV X1		0.248*		
SUPPORT	X3	0.653*		
PB	X4		0.817*	
LOC	X6		0.375*	
BUFFER X7				0.852*

* $p < 0.05$

Table 25. A Structural Equation Model With a Buffering Effect of Exogenous Variables on Caregivers' Burden (η_1).

Exogenous Latent Variable	Effect on	Caregiving Burden
Coping	γ_{11}	-0.070
Distress	γ_{12}	-0.744*
Buffering	γ_{13}	-0.338
Errors in Equation (ζ)		-0.753*

* $p < 0.05$

Table 26. Goodness of Fit (GOF) Indices

	Model 1	Model 2
Chi-square with df	27.730 (20df)	63.56 (25df)
GOF	0.937	0.883
Adjusted GOF	0.858	0.738

Notes: Model 1 with main effects only; model 2 with main effects and an interaction term.

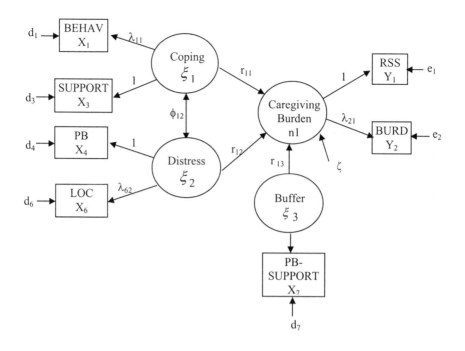

Figure 59. A Covariance Structure Model of Caregiving Burden with
an Interaction Term (Buffering Effect)

ASSESSING THE EFFECT OF INFORMATIC INTEGRATION ON HOSPITAL EFFICIENCY IN THE INTEGRATED CARE DELIVERY SYSTEMS: An Application

Following the seven steps of causal analysis outlined in Chapter 2, an empirical example of covariance structure modeling is presented below.

1. Identification and Specification of the Study Problem

It is a well-documented fact that consolidation of multiple health care organizations has increased efficiency through economies of scale in purchasing, transacting, and management. Robinson (1999) states that diversified health strategies can improve corporate performance, and that physician-hospital organizations (PHO) can enhance both clinical coordination and bargaining leverage with health plans.   However, little is known about how integrated care delivery systems (IDSs) function as system entities in applying non-structural or non-governance mechanisms such as information integration strategies. The beneficial effect of health information system integration on organizational performance has not been established (Chen and Gough, 1995; Devers, et al., 1994; Leggat and Leatt, 1997; Lin and Wan, 1999; Tan and Hanna, 1994; Van-der-Werff, 1997). The study of IDS system operations is of paramount importance, particularly to identify the best practices in system management (Wan et al., 2001). The role of health information system management in achieving high levels of hospital performance should be explored.   This study examines the effect of knowledge management through integrating clinical, administrative and managerial information systems in integrated health care delivery systems. This integration mechanism may be pivotal in the application of decision support systems in health services management; it is therefore operationally defined as health informatic integration.   System performance can be viewed from several aspects of operating efficiency: cost efficiency, bed occupancy rate, and technical efficiency; the research question is: does informatic integration improve IDSs' efficiency in hospital care?

2. Selection of an Informed Theoretical Framework

Several studies have drawn on economic and organizational theories to discuss the potential benefits of integration strategies (Conrad, 1992; Conrad and Shortell, 1996; Devers et al., 1994; Leggat and Leatt, 1997; Wang et al., 2001).   Commonly cited benefits of integration are improved clinical and administrative efficiency, and fewer unnecessary services; increases in market power, negotiation power and environmental acceptance; enhanced relationships with customers; higher profits; and improved quality of care (Brown, 1996; Conrad and Dowling, 1990; McMeekin, 2000). Because of these presumed benefits from integrated health care delivery systems, they are considered necessary for adapting to ever-increasing health care costs, aging demographics, rapid technological advancement, limited human resources, and shifts in responsibility (Fox, 1989). An IDS is viewed as an opening system that is constantly adapting to the chaotic environment.   Thus, IDSs in different contexts or environments may form a variety of integration strategies to optimize efficiency.   The contingency theory with specified relationships among context-design-performance factors is a well-suited theoretical framework for analyzing the effect of informatic integration on hospital efficiency (Shortell and Kaluzny, 1994; Bergeron et al., 2001). A covariance structure model of the effect of health informatic integration on hospital performance is formulated and validated by an empirical examination of 973 IDSs operating in the United States.

3. Quantification of the Study Variables

The American Hospital Association's Annual Survey (1998) and Dorenfest's Survey of Information Systems in Integrated Health care Delivery Systems (1998) were used. A cross-sectional study was conducted, using IDS as the unit of analysis. The study sample includes all hospital-based integrated health care delivery systems operating in the United States, from the list of Dorenfest's Survey on Information Systems in Integrated Health care Delivery Systems.

Measurement of the Study Variables. Informatic integration, an exogenous latent variable, is characterized as the use of a variety of automated application systems by an IDS to integrate its administrative (ADMSYS), management (MGTSYS) and clinical functions (CLISYS). The definitions of the study variables are presented in Table 27. System performance is viewed in this study as an IDS's efficiency – used as a neutral term denoting levels of either efficiency or inefficiency – which refers to cost, process and technical efficiency in service operations (Wan, 1995). Those are measured, respectively, by case-mix adjusted cost per hospital admissions (ACOST), unoccupied bed rate (UNOCC), and technical inefficiency score (TINEFF) obtained from the data envelopment analysis (DEA). Higher adjusted costs per hospital admission, higher unoccupied bed rate, and a higher technical inefficiency score are indicative of organizational inefficiency. Therefore, the three indicators of inefficiency indicate an IDS's performance, a latent endogenous variable.

Table 27. Definitions of the Study Variables

Variable	Label	Description	Source
Latent Variables: Exogenous: Informatic Integration			
Administration information system	ADMSYS	Number of automated application systems in the integrated care delivery system's administration function	Dorenfest
Management information system	MGTSYS	Number of automated application systems in the integrated care delivery system's management function	Dorenfest
Clinical information system	CLISYS	Number of automated application systems in the integrated care delivery system's clinical function	Dorenfest
Latent Variables: Endogenous: Hospital Inefficiency			
Cost per hospital admission	ACOST	Total cost divided by hospital admissions, adjusted by Medicare case mix index	Dorenfest
Unoccupied bed rate	UNOCC	One minus the ratio of total acute care patient days to acute care staffed beds multiplied by 365 days, adjusted by Medicare case mix index	AHA
Technical inefficiency	TINEFF	One minus technical efficiency score (IOTA). Data envelopment analysis: Input: number of aligned physicians, acute-care hospital beds Outputs: number of Medicare discharges, number of Medicaid discharges, number of inpatient surgeries, number of outpatient surgeries	AHA

DEA is a non-parametric linear programming technique that has no assumptions about the form of production. It was first introduced by Farrell (1957) and further developed by Charnes, Cooper and Rhodes (1978) and Charnes, Cooper and Lewin (1995). DEA measures relative efficiency by the ratio of total weighted output to its total weighted input (Sexton, 1978). The idea of DEA is a frontier approach to estimating technical efficiency. DEA calculations maximize the relative efficiency score of each decision-making unit (DMU). A technical definition is "the efficiency measure of a DMU is defined by its position relative to the frontier of best performance established mathematically by the ratio of weighted sum of outputs to weighted sum of inputs" (Norman & Stoker, 1991, p. 16). The inputs and outputs and any associated weights are assumed to be greater than zero. A score of one represents an efficient DMU. Because DEA is particularly well suited to multiple outputs, it has been applied in assessing hospital technical efficiency (Lynch and Ozcan, 1994; Morey et al., 1990; Bannick and Ozcan, 1995;

Ozcan, 1995).   In this case, two input variables and four output variables
were included in the DEA model. The two input variables were the number of
aligned physicians and of acute-care hospital beds. The four output variables
were the number of Medicare discharges, the number of Medicaid discharges,
the number of inpatient surgeries, and the number of outpatient surgeries.

4. Specification of the Analytical Model

AMOS (Analysis of Moment Structure) 4.0, a multivariate statistical package,
is used to validate the measurement models of the exogenous latent variable
(informatic integration) and the endogenous latent variable (inefficiency)
independently, via confirmatory factor analysis; and to test the structural
relationship between informatic integration mechanisms and inefficiency, via
covariance structure modeling of IDS performance (Figure 60).

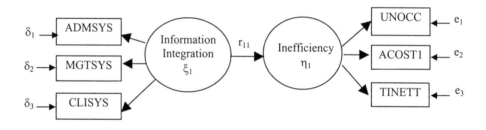

**Figure 60. A Covariance Structure Model of the Relationship Between Informatic
Integration and Hospital Performance**

5. Selection of the Intervention Design

This is not an experimental study. No randomization could be applied in the
design.

6. Confirmatory Analysis

The measurement models of informatic integration and efficiency are
separately validated by confirmatory factor analysis. Analysis of the
maximum likelihood estimates for the individual parameters and the overall

model fit is performed. The factor loadings, or standardized estimated regression coefficients in the general regression model, show moderate and statistically significant relationships between the observed variables and their corresponding latent construct. The factor loadings for CLISYS, ADMSYS, and MGTSYS are 0.83, 0.65, and 0.51, respectively, with statistical significance at the 0.05 level (Figure 61). For the measurement model of the endogenous latent variable, IDS inefficiency in hospital care, the three observable indicators are the case-mix adjusted cost per hospital admission (ACOST), unoccupied bed rate (UNOCC), and technical inefficiency score (TINEFF). The results of confirmatory factor analysis show that the factor loadings for ACOST, UNOCC, and TINEFF are 0.22, 0.38, and 0.76, respectively.

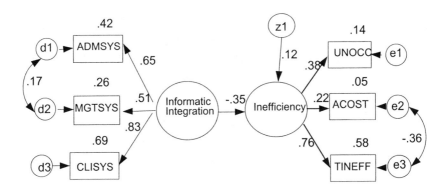

Figure 61. The Relationship between Informatic Integration and IDS Efficiency

A covariance structure analysis is then performed to determine the relationship between informatic integration and IDS inefficiency. A statistically significant relationship (-0.35) was found between these two latent variables (Table 28). The negative regression coefficient indicates that the higher the informatic integration observed among IDSs, the lower their levels of inefficiency. Informatic integration accounts for 12.2% of the variation in IDS inefficiency. The goodness-of-fit statistics of the model show that the model fits reasonably well.

7. Establishment of Causality

There is an inverse relationship between informatic integration and inefficiency. The beneficial effect of informatic integration has been gradually realized. High efficiency in hospital care can be achieved by applying decision support systems in clinical, administrative and managerial operations (Smith, Bullers and Piland, 2000). However, it is important to note that the IDSs studied here may be in different life-cycle stages. The limited effect of informatic integration through decision support and enterprise data systems for administrative, managerial and clinical functions may indicate that most of the IDSs have just begun to reengineer information technology/information systems (IT/ISs). Given the health care sector's average investment of less than three percent in IT/IS per year, it will take time to fully realize the gain from IT/IS investment in system operations.

Table 28. The Relationship between Informatic Integration and IDS Inefficiency (N=973)

Variables		Standardized Estimates (Factor Loadings)	Unstandardized Estimates	t-values
I. Measurement Models				
Exogenous Latent Variable	Indicators			
(Informatic integration)	CLISYS	.83	1.00	
	ADMSYS	.65	.52	7.13*
	MGTSYS	.51	.21	6.78*
Endogenous Latent Variable				
(Inefficiency)	ACOST	.22	129.81	2.44*
	UNOCC	.38	1.00	
	TINEFF	.76	1.54	3.67*
II. Structural Equation Model: Effect On Inefficiency				
		Standardized Estimates (γ^{**})	Unstandardized Estimates	t-value
Exogenous Variables				
Informatic Integration		-0.35	-0.013	-3.40*
III. Goodness-of-fit Statistics				
R-square (%)		12.2		
X^2		16.06		
Degree of freedom (df)		6		
X^2/df		2.68		
P value		0.013		
CFI		.999		
RMSEA		.042		
HOELTER (.05)		763		

* Significant at 0.05 or lower level
** A gamma coefficient or direct effect of an exogenous variable on IDS inefficiency

CONCLUSION

Covariance structure modeling is a fruitful approach to dealing with the structural relationships between theoretical constructs. The above empirical analysis only partially examines the context-design-performance framework. The next step in the validation of this theoretically specified framework is to add the direct linkages between such context variables as managed care penetration, market competition, and an IDS' organizational characteristics. This analysis illustrates the need for conceptual and methodological rigor in employing the contingency framework for health services management research.

REFERENCES

Bannick, R.R., Ozcan, Y.A. (1995). Efficiency analysis of federally funded hospitals: Comparison of Dod and VA hospitals using data envelopment analysis. *Health Services Management Research* 8 (2): 73-85.

Bergeron, F., Raymond, L., Rivard, S. (2001). Fit in strategic information technology management research: an empirical comparison of perspectives. Omega (*The International Journal of Management Science*) 29: 125-142.

Brown, M. (ed.). (1996*). Integrated Health Care Delivery: Theory, Practice, Evaluation, and Prognosis.* Gaithersburg, MD: Aspen Publishers.

Charnes, A., Cooper, W.W., Rhodes, E. (1978). Measuring the efficiency of decision-making units. *European Journal of Operational Research* 2: 429-444.

Charnes, A., Cooper, W.W., Lewin, A. (1995.). *Data Envelopment Analysis: Theory, Methodology and Applications.* (eds.). Boston: Kluwer Academic Publishers.

Chen, T.S., Gough, T.G. (1995). The design and development of a fully integrated hospital information system. *Medinfo* (Pt 1) 8: 569-573.

Chiriboga, D.A. (1982). An examination of life events as possible antecedents to change. *Journal of Gerontology* 37: 595-601.

Conrad, D.A. (1992). *Vertical Integration. In Strategic Issues in Health Care: Point and Counterpoint,* edited by W. J. Duncan, P. Ginter, and L. Swayne. Boston: PWS Kent.

Conrad, D.A., Dowling, W.A. (1990). Vertical integration in health services: Theory and management implications. *Health Care Management Review* 15 (4): 15-22.

Conrad, D.A., Shortell, S.M. (1996). Integrated Health Systems: Promise and Performance. *Frontiers of Health Service Management* 13 (1): 3-40.

Devers, K.J., Shortell, S.M., Gillies, R.R., Anderson, D.A., Mitchell, J.B., Morgan Erickson, K. L. (1994). Implementing Organized Delivery Systems: an Integration Scorecard. *Health Care Management Review* 19 (3): 7-20.

Derogatis, L.R., Cleary, P.A. (1977). Factorial invariance across gender for the primary symptom dimensions of the SCL-90. *British Journal of Social and Clinical Psychology* 16: 347-356.

Farrell, M.J. (1957). The measurement of productive efficiency. *Journal of the Royal Statistical Society* 120: 253-281.

Fox, W.L. (1989). Vertical integration strategies: More promising than diversification. *Health Care Management Review* 14 (3): 49-56.

Harkins, S.W., Price, D.D., Martelli, M. (1986). Effects of age on pain perception: thermonociception. *Journal of Gerontology* 41: 58-63.

Leggat, S.G., Leatt, P. (1997). A framework for assessing the performance of integrated health delivery systems. *Health Management Forum* 10 (1): 11-18.

Lin, B.Y.J., Wan, T.T.H. (1999). Analysis of integrated health care networks' performance: A contingency-strategic management perspective. *Journal of Medical Systems* 23 (6): 477-495.

Lin, N., Dean, A. (1984). Social support and depression: a panel study. *Social Psychiatry* 19: 83-91.

Lynch, J., & Ozcan, Y. (1994). U.S. hospital closures: An efficiency analysis. *Hospital and Health Services Administration* 39 (2): 205-220.

McMeekin, J.C. (2000). Integrated delivery systems: A possible cure for the ills of health care. *The Journal of Science and Health Policy* 1 (1): 83-86.

Morey, R.C., Fine, D.J., Loree, S.W. (1990). Comparing the allocative efficiencies of hospitals. *OMEGA International Journal of Management Science* 18 (1): 71-83.

Norman, M., Stoker, B. (1991). *Data Envelopment Analysis: The Assessment of Performance.* New York: Wiley.

Ozcan, Y. (1995). Efficiency of hospital service production in local markets: The balance sheet of U.S. medical armament. *Socio-Economic Planning Sciences* 29 (2): 139-150.

Pfeiffer, E. A short, portable mental status questionnaire for the assessment of organic brain deficit in elderly patients. *Journal of American Geriatric Society* 23:433.

Radloff, L.S. (1977). The CES-D scale: A self-report depression scale for research in the general population. *Applied Psychological Measurement* 1: 385-401.

Robinson, J.C. (1999). *The Corporate Practice of Medicine: Competition and Innovation in Physician Organization.* Berkeley, CA: University of California Press.

Sexton, T.R. (1978). The methodology of data envelopment analysis. In R.H. Silkman (Ed.), *Measuring Efficiency: An Assessment of Data Envelopment.* San Francisco: Jossey-Bass, Inc.

Shortell, S.M., Kaluzny, A.D. (1994). Organization theory and health services management. In *Health Care Management: Organization Design and Behavior*, edited by S. M. Shortell & A. D. Kaluzny (3rd ed.), Chapter 1. N.Y.: Delmar Publishers Inc.

Smith, H.L., Bullers, W.I., Piland, N.F. (2000). Does information technology make a difference in health care organization performance? A multiyear study. Hospital *Topics: Research and Perspectives on Health care* 78(2): 13-22.

Tan, J.K.H., Hanna, J. (1994). Integrating health care with information technology: Knitting patient information through networking. *Health Care Management Review* 19(2): 72-80.

Van-der-Werff, A. (1997). Organizing health systems for better care and performance by open information technology. *Studies in Health Technology and Informatics* 43 (Part A): 123-127.

Wan, T.T.H., Livieratos, B.B. (1978). Interpreting a general index of subjective well-being. *Milbank Memorial Fund Quarterly* 56: 531-556.

Wan, T.T.H. (1985). *Well-Being for the Elderly: Primary Preventive Strategies.* Lexington, Massachusetts: Lexington Books.

Wan, T.T.H. (1995). *Analysis and Evaluation of Health Care Systems: An Integrated Approach to Managerial Decision Making.* Baltimore: Health Professions Press, Inc.

Wan, T.T.H., Lin, Y.J., Ma, A. (2001). Integration mechanisms and hospital efficiency in integrated health care delivery systems. *Journal of Medical Systems.* (In press).

Wang, B.B., Wan, T.T.H., Clement, J., Begun, J. (2001). Managed care, vertical integration strategies and hospital performance. *Health Care Management Science* 4: 181-191.

Ware, J.W., Sherbourne, C.D. (1992). The MOS 36-Item Short-Form Health Survey (SF-36): I. Conceptual Framework and Item Selection. *Medical Care* 30 (6): 473-483.

CHAPTER 10

MULTIPLE GROUP COMPARISON WITH PANEL DATA

INTRODUCTION

After building the general structural equation model, the stability of the model will be tested over time. This chapter introduces panel analysis of repeated measures, or longitudinal data. A panel design is a longitudinal study of the same individuals over time. This design is very useful for comparing a number of treatment and control groups, regardless of whether individuals have been assigned randomly to the groups or not. Causal relationships among the study variables may be more appropriately detected in a panel design than in a cross-sectional design. Working with the same observations (i.e. organizations), the effects of many exogenous variables in the analysis can be statistically controlled. Thus, the structural relationships between the endogenous and the exogenous variables become more clearly delineated in a longitudinal study. The LISREL model can be used to analyze panel data with equality constraints applied (Table 29). The detailed description of hypothesis testing can be found in the LISREL manual (Jöreskog and Sörbom, 1993).

Table 29. Hypothesis Testing of Equality Constraints

	(1)	(2)		(G)
Measurement	$\Lambda_y =$	$\Lambda_y =$	....	$= \Lambda_y$
Measurement	$\Lambda_x =$	$\Lambda_x =$	....	$= \Lambda_y$
Measurement	$\Theta_\varepsilon =$	$\Theta_\varepsilon =$	....	$= \Theta_\varepsilon$
Measurement	$\Theta_\delta =$	$\Theta_\delta =$	....	$= \Theta_\delta$
Structural Relations	$\Phi =$	$\Phi =$	....	$= \Phi$
Structural Relations	$\Psi =$	$\Psi =$	....	$= \Psi$
Structural Relations	$B =$	$B =$	....	$= B$
Structural Relations	$\Gamma =$	$\Gamma =$	....	$= \Gamma$

This chapter was prepared by Thomas T.H. Wan, Barbara A. Mark, and Marie Gerardo.

Specifications and examples that illustrate the overall analysis with two-wave data are presented below.

Equality Constraints for Panel Data

Principles

In performing LISREL analysis, investigators should make specific assumptions about the equality of lambda matrices (LX and LY) with respect to the measurement models of the exogenous variables (X's) and endogenous variables (Y's) measured at multiple time points. For example, Figure 62 presents physical well being (η) reflected by three key indicators that were measured at T_1 and T_2. Each of the three indicators of physical health was measured twice. The measurement error (ε) associated with each indicator was identified. The linkage between the latent construct (η) and the indicator is a lambda (λ) or factor loading. Equality constraints were assumed for the respective lambdas: 1) $\lambda_{11} = \lambda_{42}$; 2) $\lambda_{21} = \lambda_{52}$; and 3) $\lambda_{31} = \lambda_{62}$.

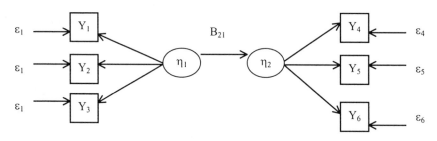

Figure 62. A Measurement Model of Physical Well-Being with Panel Data

The LISREL setup statements for the above model are as follows:
```
MO NY=6 NE=2 LY=FU,FI BE=SY,FI PS=SY,FI TE=FU,FI
FR LY 2,1 LY 3,1 LY 5,2 LY 6,2 BE 2,1
ST 1 LY 1,1 LY 4,2
FR TE 1,1 TE 2,2 TE 3,3 TE 5,5 TE 6,5
EQ LY 1,1 LY 4,2
EQ LY 2,1 LY 5,2
EQ LY 2,1 LY 1,3 LY 6,2
OU ALL
```

An Example

The following example illustrates the measurement model of patient satisfaction with the two waves of panel data. Patient satisfaction has long been used as a key indicator of the quality of care. Although the multidimensional measure of patient satisfaction is believed to be reliable, the use of an aggregate overall summary index has not been verified as appropriate (Boles and Wan, 1992). The cross-sectional data on patient satisfaction do not provide direct evidence of the stability of the measurement. Thus a study is needed to examine whether the stability of an aggregate summary measure of patient satisfaction is lower than the stability of the separate summary measures such as satisfaction with accessibility, availability, technical care, art of care, cost of care, etc. Ware and associates (1978) identified eight dimensions and measurement variables of patient satisfaction as follows:

Art of Care: Concern, consideration, friendliness, patience, and sincerity.

Technical Care: Competence, adherence to high standards, ability, accuracy, experience, thoroughness, attention to detail, good examination, and adequate information.

Accessibility/Convenience: Time and effort required to get an appointment, distance/proximity to site of care, time and effort required to get to site of care, hours care is available, availability of assistance over telephone, and availability of home care.

Finance: FFS costs, PPS costs, flexibility of payment mechanisms, and comprehensive insurance.

Physical Environment: Pleasant atmosphere, comfort of seating, attractive waiting rooms, clarity of signs/directions, good lighting, quiet, clean, neat and orderly facilities/equipment.

Availability: Adequate supply of physicians, nurses and other providers, adequate supply of clinics and hospitals.

Continuity: Regularity of care from the same facility, location and provider.

Efficacy/Outcomes: Perceptions/ beliefs that the physicians cure, relieve suffering, and prevent diseases.

A panel design was used to collect data from 1,451 Medicare beneficiaries who had had previous usual sources of care before enrollment in risk-based HMOs (Ho, Stegall and Wan, 1994; Rossiter, Wan, and Langwell, 1988). Patient satisfaction was measured by two dimensions: (1) Satisfaction with quality of care, and (2) satisfaction with access to care. Indicators of the latent endogenous variables are presented in Table 30.

Table 30. Indicators of Patient Satisfaction in the Ambulatory Care Patient Population

Dimension	Indicators	Definition
Quality of Care (QOC)	PROCARE	Professional competence of providers
Quality of Care (QOC)	DISCUSS	Physician's willingness to discuss or explain medical problems
Quality of Care(QOC)	COURT	Staff courtesy & consideration
Quality of Care (QOC)	OVERSAT	Overall satisfaction
Access to Care (ATC)	COVENT	Traveling to the clinic
Access to Care (ATC)	APPT	Convenience of appointment
Access to Care (ATC)	WAIT	Length of waiting time in the office

Patient satisfaction was conceived with two related latent constructs (quality of care [QOC] and access to care [ATC]). Separate measurement models for quality of care and access to care were developed for the two-wave data and are presented in Figure 63 and Figure 64, respectively. In addition, a cross-lagged model of QOC and ATC in the panel study is presented in Figure 65. This model assumes that QOC at Time 1 (T_1) affects ATC at Time 2 (T_2) and ATC at Time 1 affects QOC at Time 2.

The results show that for two separate summary measures of patient satisfaction, stability was relatively low (0.305 for QOC and 0.354 for ATC).

The stability of an aggregate summary measure of patient satisfaction was lower than that of the two separate summary measures (0.211 for QOC and 0.299 for ATC). Since moderate causal relationships and cross-lagged effects (the measure of T_1 affecting that of T_2) between the two latent constructs were found, both constructs not only affect each other, but also may have something in common shared by the indicators (e.g., DISCUSS and WAIT). This study concludes that due to the low stability and relative influences between the two dimensions, an overall measure obtained by summing all the item scores is not appropriate. Therefore, separate measures with summary scores are recommended for measuring patient satisfaction.

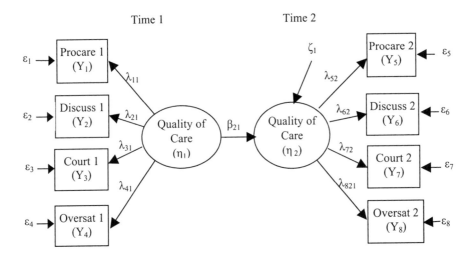

Figure 63. Two-Wave Panel Study of Patient Satisfaction with Quality of Care

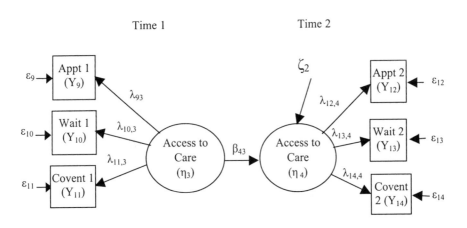

Figure 64. Two-Wave Panel Study of Patient Satisfaction with Access of Care

Time 1   Time 2

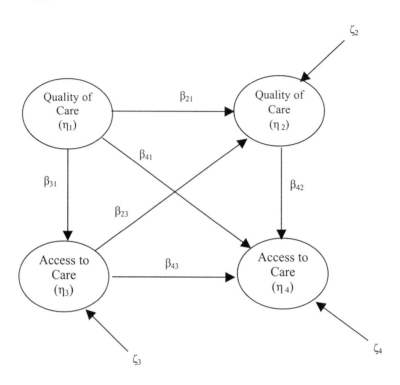

Figure 65. Two Dimensions of Patient Satisfaction: A Panel Study

Another Example: A Study of Physical Health Over Time

Wan (1985) examined critical life change events such as retirement (RET) and widowhood (WIDOW) for their effects on the physical health of the elderly, at two time points, with age, income and previous physician visits simultaneously controlled, in a structural equation model (Figure 66). The stability of physical health, measured by three indicators, is indicated by β_{21}. The study examined the effects of retirement and widowhood on health, measured according to the timing of the events (e.g., number of months before the measurement of physical health). Interestingly enough, widowhood exerted a latency effect on poor health at Time one, whereas retirement exerted a latency effect on poor health at Time two. Thus the panel study design examined the health trajectories affected by two critical life-change events. Readers who are interested in the long-term dynamics of health trajectories among black and white adults can review Ferraro and associates' article (1997), in which multiple group analysis is performed.

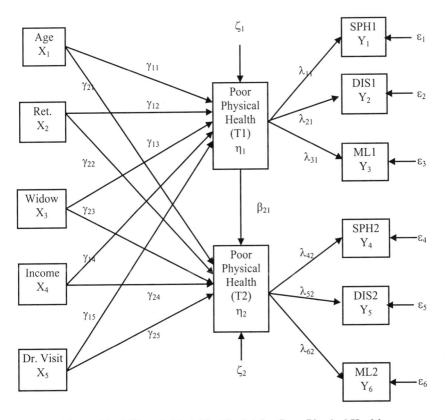

Figure 66. A Generic Model for Explaining Poor Physical Health

SPH: Self-assessed Poor Health
DIS: Chronic Condition
ML: Mobility Limitation
Ret.: Recency of Retirement
Widow: Recency of Widowhood

MULTIPLE GROUP COMPARISON

Principles

The structural equation model can be applied to multiple group analysis. This approach is also referred to as a stacked model (Hayduk, 1987). The general principles are as follows:

1. Formulation of a generic theoretical model with clear specifications.
2. Replication of the same generic model in multiple subgroups.
3. Evaluation of the adequacy of the theoretical model by examining the overall GOF statistical indices.
4. If the generic model is unable to fit to the data, it is necessary to reformulate the model for a specific subgroup.
5. Significance Test of the Difference of Slopes (two groups):

To compare the effect of a given (exogenous or endogenous) variable on an endogenous variable, the regression coefficients can be compared in multiple group analysis. The following significance test is applied to two subgroups with regression estimates: $t = B_1 - B_2 / (S^2_{e1} + S^2_{e2})$, where B is an unstandardized regression coefficient and S_e is a standard error estimate. The degree of freedom is $N_1 + N_2 - 2$; N_1 and N_2 refer to the sample sizes for the respective groups.

An Application of Multiple Group Analysis: Organizational Determinants of Patient Satisfaction

Following the seven steps in causal analysis outlined in Chapter 2, an empirical example of multiple group analysis is presented below. This serves as an illustrative purpose by using a panel data set collected by the Outcomes Research in Nursing Administration Project (Mark, Salyer, and Wan, 2001).

1. Identification and Specification of the Study Problem

In the current cost-containment environment, many hospitals have changed the organization of care delivery. Common strategies to curtail hospital costs include reducing the nursing staff, increasing the use of cross-trained personnel, implementing case management, using critical paths, and introducing product-line management, among other cost-cutting moves. How, such changes have affected the quality of health care has been a pressing question for health services researchers and health policy-makers. Therefore interest has risen in measuring patient-centered outcomes such as satisfaction.

Instruments for measurement of patient satisfaction are used to assess the overall quality of care across hospitals or other medical care settings. Many investigators have reported on the reliability of the existing instruments, but the instruments' construct validity has yet to be demonstrated (Greeneich, Long, and Miller, 1992; Kangas, Kee, and McKee-Waddle, 1999; Nelson and Niederberger, 1990; Strasser et al., 1993). Health care executives are interested in how organizational factors may affect the variation in patient satisfaction. The purpose of this empirical example is to demonstrate the validity of a measurement model of patient satisfaction and to identify organizational factors that may influence patient satisfaction, using covariance structure modeling with multiple group analysis.

2. Selection of an Informed Theoretical Framework

Ware, Davies, and Stewart (1978) reviewed the literature on patient satisfaction with medical care and identified eight distinguishable dimensions: art of care, technical quality of care, accessibility/convenience, finances, physical environment, availability, continuity of care, and efficacy/outcomes. Nelson and Niederberger (1990) examined patient satisfaction and formulated six quality-related indicators: 1) access; 2) administrative management; 3) clinical management (qualifications, technical skills); 4) interpersonal management (staff warmth, explanation of care); 5) continuity of care; and 6) general satisfaction. Greeneich and colleagues (1992) identified three factors that influence patients' satisfaction with nursing care: nursing personality characteristics (friendliness, social courtesy, helpfulness, and kindness), nursing caring behaviors (empathy, compassion, communication, and comfort measures), and nursing proficiency (knowledge, technical proficiency, and organizational skills). Strasser et al. (1993) pointed out that patient satisfaction is a multidimensional construct and can be measured by a single global index of patient satisfaction. By modifying the above conceptual models, a measurement model of patient satisfaction was proposed. The proposed measurement model included a single global construct of patient satisfaction, which was measured by nine indicators (Figure 67).

Further literature review found other studies involved with the relationship between organizational factors and patient satisfaction. Bowles (1997) described the evolution of nursing information systems and the design goals for current systems. The study provided evidence of the benefits of NIS being improved efficiency, patient safety and satisfaction, and ability to measure quality. Tucker (2000) suggested that patient satisfaction differs significantly according to age, gender, education, and race.

McCloskey (1998) found an inverse relationship between registered nurse (RN) hours of care and rates of patient complaints.

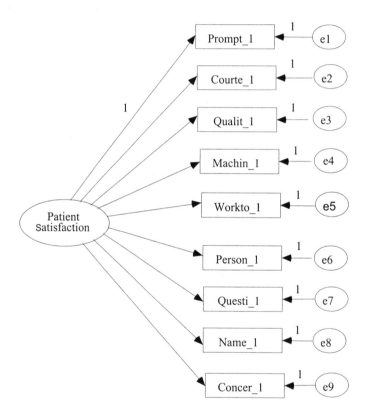

Figure 67. A Measurement Model of Patient Satisfaction

3. Quantification of the Study Variables

The data were obtained from a survey conduced by the Outcomes Research in Nursing Administration Project (Mark, Salyer, and Wan, 2001). The database includes 124 nursing units that provided data at both Time 1 and Time 2 in 62 acute care hospitals in the United States. The unit of analysis is nursing unit.

Measurement of the study variables. Nine indicators reflect patient satisfaction. Information on the nursing unit, patient characteristics, and patient satisfaction is provided in Table 31.

Table 31. Definitions of the Study Variables

Variable	Description
Patient Satisfaction	A latent endogenous variable measured by following nine indicators:
Prompt_1	Rating nurses' promptness in answering calls
Courte_1	Rating the overall courtesy and friendliness of nursing staff
Qualit_1	Rating the overall quality of nursing care
Machin_1	Rating the overall satisfaction with the equipment and machines
Workto_1	Rating the overall ability of professional staff
Person_1	Rating nurses in treating patients as persons, not diseases
Questi_1	Rating nurses' answers to patients' questions
Name_1	Rating personal care (names)
Concer_1	Rating how comfortable patients felt in voicing concerns to nurses
Exogenous Variables Include:	
Age_1	Mean age of patients
Edleve_1	Average educational level
Gender_1	% Females
Racegr_1	% Whites
Bedsize	Number of hospital beds
CMI	Case mix index
IOTA	Technical efficiency index
LICBEDS	Bed size per nursing unit
RNPROP	% RNs of the total nursing staff
HIGHTECH	Index of high tech services offered
PTDAYS	Total patient days

4. Specification in Analytical Modeling

From previous research, a covariance structure equation model was proposed (See Figure 68). This model included one endogenous latent variable, patient satisfaction, and 11 exogenous variables. These exogenous variables included Age_1, Edleve_1, Gender_1, Racegr_1, Bedsize, CMI, IOTA, LICBEDS, RNPROP, HIGHTECH, and PTDAYS.

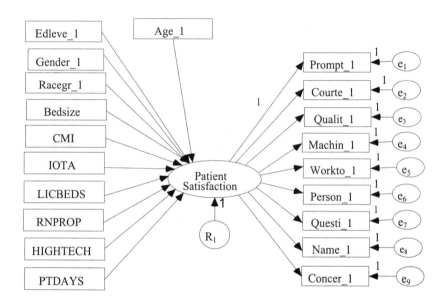

Figure 68. Covariance Structure Model of Patient Satisfaction

This study employs covariance structural analysis, which provides parameter estimates simultaneously for the measurement model and for the structural equation model, to assess the effects of various organizational factors on quality of care as indicated by patient satisfaction. The AMOS 4 computer program is the analytic tool used. In the measurement model, each indicator is considered a linear function of the common factor (latent construct). That is, the relationship among observed indicators and their latent constructs can be illustrated as: $Y = \Lambda_Y \eta + \varepsilon$, where:

Y is a vector of observed endogenous variables;

Λ_Y is a matrix of the loading of the observed Y variables on the latent η variables; and

ε is a vector of unique factors.

The specification of the structural equation model is: Patient satisfaction $=f$ (organizational characteristics) + residual terms. Mathematically, the structural equation model can be expressed as: $\eta = B\eta + \Gamma\xi + \zeta$, where:

η is a vector of latent endogenous variables;

ξ is a vector of latent exogenous variables;

B is a matrix of coefficients relating the endogenous variables one to another;

Γ is a matrix of coefficients relating the exogenous variables to the endogenous variables; and

ζ is a vector of errors in equations.

In LISREL analysis, the structural equation model defines the causal links among the exogenous latent variables factored from observed variables in the measurement model, as well as the effects of the exogenous variables. To assess the validity of the measurement model of patient satisfaction, confirmatory factor analysis (CFA) was used to assess how well nine indicators represent the endogenous latent variable. First, CFA was conducted using the total sample. Then a multiple group analysis of the measurement model was performed using equality constraints to compare two types of hospital models. The total sample was divided into two groups: large hospitals and small hospitals. The large hospital group has a bed size equal or larger than 450, and the small hospital group has a bed size less than 450. Finally, to determine the relationship between organizational factors and patient satisfaction, multiple group analysis of the covariance structure model was performed.

5. Selection of the Intervention Study Design and Analysis

Descriptive statistics and a correlation matrix of all nursing units of those in small hospitals and of those in large hospitals are presented in Tables 32, 33, and 34, respectively. All nine indicators were highly correlated, which suggested a high reliability of the instrument. However, some indicators were extremely highly correlated, suggesting that there may be some redundant items in the measurement instrument.

Table 32. Descriptive Statistics and Correlation Matrix of All Nursing Units (N = 124)

Indicators	Mean	S.D.	1	2	3	4	5	6	7	8	9
1. CONCER_1	5.14	.41	1.000								
2. COURTE_1	4.72	.56	.804	1.000							
3. MACHIN_1	4.23	.52	.675	.726	1.000						
4. PERSON_1	5.34	.51	.655	.646	.540	1.000					
5. PROMPT_1	4.11	.68	.746	.841	.764	.667	1.000				
6. QUALIT_1	4.54	.55	.784	.903	.760	.655	.863	1.000			
7. QUESTI_1	5.30	.45	.837	.733	.689	.698	.746	.789	1.000		
8. WORKTO_1	4.38	.55	.733	.795	.768	.621	.833	.855	.738	1.000	
9. NAME_1	5.20	.48	.541	.469	.431	.449	.490	.488	.569	.437	1.000

Note: all correlation coefficients are significant at the 0.01 level (2-tailed).

Table 33. Descriptive Statistics and Correlation Matrix of Nursing Units from Small Hospitals (N = 62)

Indicators	Mean	S.D.	1	2	3	4	5	6	7	8	9
1. CONCER_1	5.13	.42	1.000								
2. COURTE_1	4.74	.52	.767	1.000							
3. MACHIN_1	4.23	.56	.629	.734	1.000						
4. PERSON_1	5.26	.53	.644	.613	.533	1.000					
5. PROMPT_1	4.12	.73	.741	.863	.790	.700	1.000				
6. QUALIT_1	4.57	.54	.759	.873	.793	.626	.859	1.000			
7. QUESTI_1	5.31	.49	.860	.758	.705	.677	.775	.828	1.000		
8. WORKTO_1	4.37	.59	.723	.784	.819	.608	.824	.876	.768	1.000	
9. NAME_1	5.20	.48	.575	.411	.378	.413	.495	.431	.531	.403	1.000

Note: all correlation coefficients are significant at the 0.01 level (2-tailed).

Table 34. Descriptive Statistics and Correlation Matrix of Nursing Units from Large Hospitals (N=62)

Indicators	Mean	S.D.	1	2	3	4	5	6	7	8	9
1. CONCER_1	5.14	.40	1.000								
2. COURTE_1	4.70	.61	.855	1.000							
3. MACHIN_1	4.23	.48	.748	.737	1.000						
4. PERSON_1	5.45	.46	.699	.744	.577	1.000					
5. PROMPT_1	4.11	.63	.756	.835	.720	.648	1.000				
6. QUALIT_1	4.51	.58	.822	.936	.730	.759	.881	1.000			
7. QUESTI_1	5.29	.42	.807	.721	.661	.780	.697	.747	1.000		
8. WORKTO_1	4.39	.51	.749	.831	.680	.666	.847	.842	.690	1.000	
9. NAME_1	5.20	.49	.498	.534	.511	.524	.487	.556	.629	.487	1.000

Note: all correlation coefficients are significant at the 0.01 level (2-tailed).

6. Confirmatory Analysis

As described earlier, the generic measurement model of patient satisfaction included nine indicators (Figure 69). Prior to analysis of the covariance structure model, the measurement model was analyzed. The summary statistics of the measurement model can be found in Table 35. The standardized coefficients of all the indicators were high and statistically significant. The estimated measurement errors were also statistically significant. However, the X^2 likelihood ratio (X^2/df) for this model was 2.797 with p = .000, the adjusted goodness of fit was .775, and the Hoelter index (.05) was 59. These statistics indicate that the generic measurement model fits poorly to the data, and that modification of the model is needed.

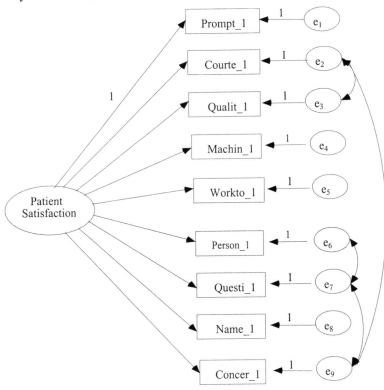

Figure 69. Revised Measurement Model of Patient Satisfaction

Table 35. Summary Statistics for the Measurement Models of Patient Satisfaction, with and without Correlated Measurement Error

Estimate	Without correlated measurement error	With correlated measurement error
	Standardized Coefficient	Standardized Coefficient
Y_1	.911	.921
Y_2	.922	.895
Y_3	.949	.938
Y_4	.811	.823
Y_5	.889	.903
Y_6	.718	.710
Y_7	.845	.837
Y_8	.542	.538
Y_9	.854	.825
Error	Estimate	Estimate
e_1	.079	.071
e_2	.047	.062
e_3	.030	.036
e_4	.093	.088
e_5	.063	.056
e_6	.124	.126
e_7	.059	.061
e_8	.162	.163
e_9	.045	.054
e_{79}	---	.468
e_{23}	---	.408
e_{29}	---	.272
e_{67}	---	.201
Goodness-of-fit Statistics		
X^2	75.515	24.605
Degrees of Freedom (df) 27		23
P	.000	.371
X^2/df	2.797	1.070
GFI	.865	.952
AGFI	.775	.906
RMSEA	.128	.025
Hoelter (.05)	59	158

* Statistically significant at .05 level

A revised measurement model was formulated to improve its fit to the data by introducing correlated measurement errors. The revised measurement model is depicted by the diagram shown in Figure 70. Table 35 also shows lambda's values for the revised measurement model. The X^2 likelihood ratio (X^2/df) for this model was 1.07 with p = .371, the adjusted goodness of fit was .906, and the Hoelter index (.05) increased to 158. Therefore, the fit of this model was very good. However, if we want to look at the applicability of this model, a multi-group (stacked model) study is necessary. Here, we want to examine whether the revised model remains true for subgroups of hospitals, large vs. small. The stacked model is simply a replicated model for the two subgroups, but with equality constraints applied to Λy matrices (Figure 70). The results of the measurement model for the stacked group are illustrated in Table 36. The X^2 likelihood ratio (X^2/df) for this model was 1.363 with p = .039, and the adjusted goodness of fit was 0.79. The results indicate that the measurement model did not fit well for either subgroup. However, given the small sample size of the subgroups and the good model fit for the whole sample, no further revision is necessary for the measurement model.

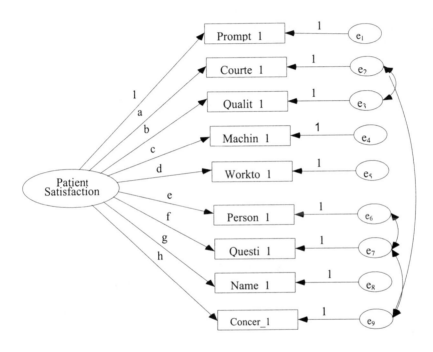

Figure 70. Measurement Model of Patient Satisfaction with Multiple Group Analysis (Small vs. Large Hospital Size)

Table 36. Goodness-of-fit Statistics of Stacked Model

GOF statistics	Estimates
X^2	73.612
Degrees of freedom	54
P-value	0.039
X^2/df	1.363
GFI	.874
AGFI	.790
RMSEA	.058
Hoelter (.05)	108

After the measurement model of patient satisfaction was validated, the covariance structure model was analyzed. The covariance structure model is depicted by the diagram shown in Figure 71. The summary statistics for the structure equation model are presented in Table 37. The results showed that only three exogenous variables were statistically significant. These were Age_1, Hightech, and Ptdays. The GOF statistics showed that the model had a poor fit, P = .000, X^2/df was 3.019, and the adjusted GFI was .65. Those statistics suggested that this model needed to be revised.

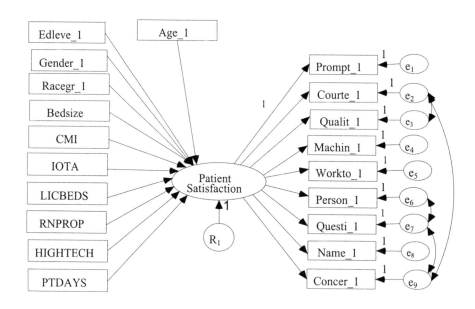

Figure 71. Covariance Structure Model of Patient Satisfaction with Correlated Errors

Table 37. Summary Statistics for the Generic Structural Equation Model and Revised Model

Predictors	Generic Model Standardized Estimate	Revised Model Standardized Estimate
Age_1	.203*	.227*
Hightech	.226*	.113
Ptdays	-.251*	-.317*
Edleve_1	-.071	
Gender_1	.025	
Racegr_1	.140	
Bedsize	-.077	
CMI	-.165	
IOTA	-.118	
Licbeds	-.047	
Rnprop	.037	
GOF statistics		
X^2	501.136	56.098
Df	166	50
P-value	.0000	.257
X^2/df	3.019	1.122
GFI	.724	.920
AGFI	.650	.875
RMSEA	.135	.033
Hoelter (.05)	44	133

* Statistically significant at .05 level

The revised structure equation model was conducted by eliminating the statistically insignificant exogenous variables. The revised model is illustrated in Figure 72. For this model, age_ 1 and Ptdays were significant, and the model fitted fairly well. The X^2 likelihood ratio (X^2/df) for this model was 1.122, with p = .257 and the adjusted goodness of fit was .875. The Hoelter index (.05) increased from 44 to 133.

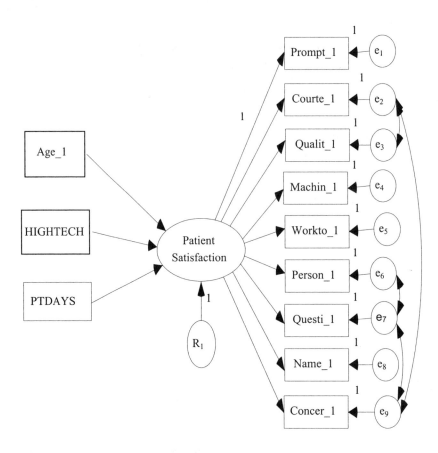

Figure 72. Revised Covariance Structural Model of Patient Satisfaction

7. Establishment of Causality

In general, this study has validated the measurement model of patient satisfaction with nine indicators. The correlated measurement errors suggest that these indicators have shared common variance, not accounted for by the construct. The proposed structural equation model of the organizational determinants of patient satisfaction does not fit well to the sample data. Only three organizational indicators were found to significantly predict the variation in patient satisfaction. The revised model fits better to the data. The findings on the lack of effects of organizational factors on patient satisfaction are supported by other studies. For example, Kangas et al. (1999) explored differences and relationships among the job satisfaction of registered nurses, patient satisfaction with nursing care, nursing care delivery models, organizational structure, and organizational culture. There were no differences in nurses' job satisfaction or in patient satisfaction with nursing care in different organizational structures or where different nursing care delivery models were used.

Further studies may include more organizational characteristics of hospitals or nursing units. The limitation of this study is that patient satisfaction was treated as a single theoretical construct. In the future studies, multiple dimensions of the patient satisfaction construct should be considered.

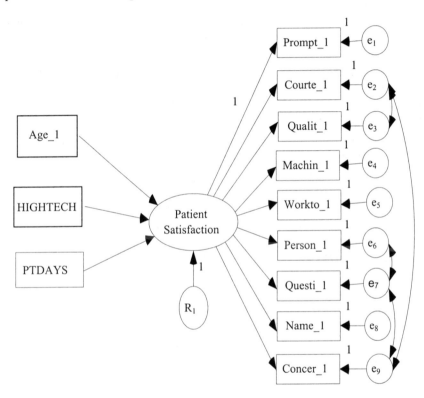

Figure 73. Revised Covariance Structural Model of Patient Satisfaction

SUMMARY AND CONCLUSION

When the same variables are used over time, there is a tendency for the corresponding error terms to be correlated. Therefore, it is necessary to specify the intercorrelations between the error terms. In addition, other unknown variables or factors may affect the question of interest. Factors such as changing health behavior or attitudes may confound the study results.

In a panel study, missing values due to attrition or non-response may cause a serious problem for multiple group analysis. If a list-wise deletion of observations is applied, a substantial reduction of the study subjects or observations may occur. Thus results from the analysis of the remaining observations may not be generalized to the original sample. An alternative strategy is to use imputed values (e.g., means or regression estimates) for the missing cases so that the analysis can be performed on the original sample. Statistical methods are available from AMOS to generate maximum likelihood estimates of missing values for the study variables. However, it is highly recommended that the investigators perform the model fit with and without missing cases, so that comparable results in the analyses can be identified and evaluated.

Multiple group analysis is a useful analytical method in structural equation modeling. Both short-term and long-term dynamics in organizational performance can be examined with a panel study design. For example, the organizational and market determinants of hospital quality of care and hospital efficiency can be analyzed in two different types of hospitals (profit vs. non-profit). If evidence shows that organizational strategies such as clinical and informatic integration and technological emphasis, and market forces such as managed care penetration and competition affect not only hospitals' efficiency, but also the quality of their care, this structural equation model can be validated by multiple group analysis with both profit and non-profit hospitals. The relative fit of this model can be judged by a variety of measures (Browne and Cudek, 1993; Jöreskog and Sörbom, 1993), as noted in previous chapters. When the GOF indices indicate that the model does not fit the data, it is necessary to trim or reduce the complexity of the model to derive a model with better fit. However, when no comparable results are generated in multiple group analysis as one investigates the performance differences between profit and non-profit hospitals, separate theoretical specifications are called for.

REFERENCES

Boles, M., Wan, T.T.H. (1992). Longitudinal analysis of patient satisfaction among Medicare beneficiaries in different HMOs and fee-for-service care. *Health Services Management Research* 5: 198-206.

Bowles, K.H. (1997). The barriers and benefits of nursing information systems. *Computers in Nursing* 15 (4): 191-6.

Browne, M.W., Cudeck, R. (1993). Alternative ways of assessing model fit. In Bollen, K.A. and Long, J.S. (eds.), *Testing Structural Equation Models*. Newbury Park, CA: Sage Publications.

Ferraro, K.F., Farmer, M.M., Wybraniec, J.A. (1997). Health trajectories: long-term dynamics among black and white adults. *Journal of Health and Social Behavior* 38:38-53.

Hayduk, L.A. (1987). *Structural Equation Modeling with LISREL*. Baltimore: Johns Hopkins University Press.

Greeneich, D., Long, C., Miller, S. (1992). Patient satisfaction update: Research applied to practice. *Applied Nursing Research* 8: 43-47.

Jöreskog, K.G., Sörbom, D. (1993). *LISREL 8: Structural Equation Modeling with the SIMPLIS Command Language*. Mooresville, IN: Scientific Software.

Kangas. S., Kee. C.C., McKee-Waddle. R. (1999). Organizational factors, nurses' job satisfaction, and patient satisfaction with nursing care. *The Journal of Nursing Administration* 29 (1): 32-42.

McCloskey, J.M. (1998). Nurse staffing and patient outcomes. Nursing Outlook 46(5): 199-200.

Nelson, C.W., Niederberger, J. (1990). Patient satisfaction surveys: An opportunity for total quality improvement. *Hospital & Health Services Administration* 35: 409-427.

Strasser, S. et al. (1993). The patient satisfaction process: Moving toward a comprehensive model. *Medical Care Review* 50(2): 219-248.

Tucker, J.L. (2000). The influence of patient sociodemographic characteristics on patient satisfaction. *Military Medicine* 165 (1): 72-76.

Wan, T.T.H. (1985). *Well-Being for the Elderly: Primary Preventive Strategies*. Lexington, MA: Lexington Books.

Ware, J.E., Davies, A.R., Stewart, A.L. (1978). The measurement and meaning of patient satisfaction. *Health & Medical Care Services Review* 1 (1): 3-15.

CHAPTER 11

MULTILEVEL COVARIANCE MODELING

INTRODUCTION

Health services researchers have to deal with the complex nature of hierarchical data compiled from administrative and clinical data collection systems. The records of patients in health care facilities offer multiple measurements of their physical health and functional outcomes. These outcomes are directly influenced by patients' personal characteristics or risk factors. Statistical analysis of the effects of patients' personal factors on patient outcomes is considered a micro-level analysis, or individual-level analysis. If an investigator is interested in understanding the effects on patient outcomes of such organizational factors as size, complexity, staffing, and provider characteristics, then personal data are aggregated at the organizational level. Thus, the individual-level analysis is nested within the organizational units. In multivariate statistical analysis, the hierarchical structure of personal and organizational levels of data must be carefully considered. Otherwise, the effects of individual and of organizational characteristics on patient outcomes may be inappropriately estimated.

When multi-level data are used in a study, researchers tend to disaggregate the aggregated data to the lower level, i.e. to assign the value of aggregated data to the lower level. For example, the study of nurses' performance uses hospital attributes as predictor variables. Similarly, researchers aggregate the lower (individual) level data to the upper (hospital) level by computing the means or medians of the individual-level measures, to complement the need for analyzing the performance of hospitals. From the methodological viewpoint, the former cannot satisfy the assumption of the independence of observations that underlies the traditional statistical approach (Bryce & Raudenbush, 1992; Heck and Thomas, 2000). Another problem posed by disaggregation is that statistical tests involving the variable at the upper-level unit are based on the total number of lower-level units, which can influence estimates of the standard errors and the associated statistical inference (Bryk & Raudenbush, 1992).

This chapter was prepared by Thomas T.H. Wan and Barbara A. Mark.

Small area analysis measures and compares the use of medical care services by defined geographic areas or medical market areas (Wan, 1995). The aggregate or ecological analysis of health services or health care outcomes focuses on the comparison of geographic areas rather than individuals. The aggregation may lose valuable information, in that it ignores the meaningful lower level variance in the outcome measure (Morgenstern, 1998). It may cause the "ecological fallacy," i.e. analyzing upper level data, but interpreting the results at the lower level. In fact, most data are hierarchical: for example, individuals are nested in families, families are part of communities, and communities are nested in counties and/or states. The hierarchical nature of data should not be neglected in either theory building or data analysis (Little, Schnabel, and Baumert, 2000; Muthén, 1991).

STATISTICAL SPECIFICATION OF A MULTILEVEL MODEL WITH OBSERVED INDICATORS

A special feature of multilevel modeling is using the intercept and slope as outcomes (Bryk and Raudenbush, 1992). This is illustrated in the relationship between nurses' professional practice (PROFPRAC) and job satisfaction (SATISF) at the nursing unit level:

$SATISF_{ig} = B_{0g} + B_{1g} (PROFPRAC) + r_i$.

SATISF (Y) of ith unit in the gth hospital is influenced by PROFPRAC (X). B_{0g} is the expected level of staff satisfaction, with a value of zero on the predictor, i.e. PROFPRAC. B_{1g} is PROFPRAC's slope for hospital g, and r_{ig} is the error term.

If the hospital-level variable, bed size (W), is believed to have an impact on SATISF (Y) through the intercept and slope, these relationships can then be illustrated in a simple two-level model.

Hierarchical Form:
  Level 1 (e.g., nursing units): $Y_{ig} = B_{0g} + B_{1g} X_{ig} + r_{ig.}$
  Level 2 (e.g., hospitals): $B_{0g} = r_{00} + r_{01}$ (bed size) $+ u_{0g}$,
                                 $B_{1g} = r_{10} + r_{11}$ (bed size) $+ u_{1g.}$

After replacing the intercept and slope with corresponding elements in the latter two equations, we obtain a combined model with two levels in the following equation:

SATISF$_{ig}$ = r$_{00}$ + r$_{01}$ (PROFPRAC) + r$_{10}$ (bed size) +

r$_{11}$ (PROFPRAC) (bed size) + u$_{1g}$ (bed size) + u$_{0g}$ + r$_{ig}$,

where:

r$_{00}$ is the mean effect of staff satisfaction for a hospital;

r$_{01}$ is the effect of hospital size, r$_{10}$ is the average bed size slope for a hospital; and

r$_{11}$ is the average difference in bed size slope.

The term of u$_{0g}$ is the unique effect of hospital g on average SATISF given by hospital size, the term of u$_{1g}$ is the unique effect of hospital g on the PROFPRAC's slope conditioned on bed size. r$_{00}$, r$_{01}$, r$_{10}$, and r$_{11}$ are Level-2 variance-covariance components. The variance of r$_{ig}$ is σ^2. The cluster effect (ρ) of nursing units is measured by the proportion of variance in SATISF that is between hospitals (e.g., Level 2). This coefficient, referred to as the intraclass correlation coefficient, is computed by:

ρ = r$_{00}$/ (σ^2 + r$_{00}$).

This two-level model can be easily extended to a three-level model by including "market area" as the third level of predictor variables. The hospital-level variation in an outcome variable such as SATISF is then nested in the market area studied. Adding a market characteristic such as hospital market competition into the equation will further complicate the statistical analysis of the effects of three-level indicators on nursing unit performance.

MULTILEVEL ANALYSIS WITH LATENT VARIABLES

The investigation of effects of a complex set of individual, aggregate, and contextual variables on administrative and patient outcomes is not a simple matter, particularly when multiple theoretical or latent constructs are involved in the causal analysis. The three objectives of a causal inquiry in multilevel analysis are: 1) to estimate the individual or biological effects of personal attributes on several outcome variables; 2) to estimate the ecological effects of aggregate measures of these personal attributes on the outcomes; and 3) to investigate simultaneously the contextual effects of organizational or environmental factors on the outcomes.

To illustrate causal inquiry using multi-level analysis, a two-level analysis of nursing units' performance in administrative and patient care outcomes, with a balanced design (equal number of nursing units selected from a sample of hospitals) is presented. The focus of this presentation is on the logic of theoretical specifications and the application of a comprehensive modeling program, Mplus (Muthén and Muthén, 1998). Readers should follow the scientific steps for causal inference outlined in Chapter 2 for performing a confirmatory analysis of the determinants of job satisfaction and patient satisfaction of nursing units in multiple hospitals.

1. Identification and Specification of the Study Problem

Increasingly, patient perceptions are used to measure the quality of health care in a variety of health care delivery systems. Several forces have prompted the use of patient satisfaction as an indicator of the quality of health care. These factors include increasing competition among health care providers, the emergence of managed care, emphasis on streamlining the structure and functions of nursing staff, and concern that cost containment policies may adversely affect access to care and the quality of care.

The literature on patient satisfaction at the nursing unit level identifies several reasons for using patient perception as an indicator of the quality of care. First, patient perceptions may be more sensitive to differences across health care delivery systems than are statistical measures of quality (e.g., risk-adjusted mortality). Second, unlike adverse outcome measures of care, patient perceptions can capture positive aspects of care (Rosenthal and Shannon, 1997; Shelton, 2000). Third, the principle of autonomy dictates that competent patients have the right to decide what is best for them (Vuori, 1987). Fourth, high satisfaction with care is considered a component of health status itself (Donabedian, 1980). Studies show that satisfied patients are more likely to continue using medical care (Thomas and Penchansky, 1984), to maintain a relationship with a specific provider (Marquis, 1983), and to comply with medical regimens (Dimatteor and DiNicola, 1983). Yet, despite the evidence for and the general agreement on using patient satisfaction to assess the quality of care, knowledge about the aspects of care that lead to patient satisfaction is still limited. Because of the multidimensional nature of patient satisfaction (a latent construct), it is important to ascertain the validity, reliability and applicability of a measurement instrument that can capture patient perceptions of the quality of care at the nursing unit level. Furthermore, it is important to identify how nursing professional practice and job satisfaction influence patient satisfaction, while simultaneously considering other nursing and hospital contextual factors in the analysis.

2. Quantification of the Study Variables

According to Mark, Salyer, and Wan (2001), nursing professional practice was treated as an endogenous latent variable. However, for illustrating the application of multilevel analysis, job satisfaction, and patient satisfaction were considered as endogenous latent variables. Several observed indicators could measure each of these theoretical constructs. Because the unit of analysis is the nursing unit, patient satisfaction measures and nursing attributes should be aggregated first to the nursing unit level and then to the hospital level. Thus, a two-level analysis of covariance structure model is designed. The measurement models of the three latent variables are summarized in Table 38.

Table 38. Latent Constructs and Their Indicators

Latent Construct	Indicator	Description
Patient Satisfaction		
	Prompt	Rating promptness in answering calls
	Courtesy	Rating the overall courtesy and friendliness of staff
	Quality	Rating the overall quality of nursing care
	Machine	Rating the overall satisfaction with the equipment and machines
	Workto	Rating the overall ability of professional staff
	Person	Rating nurses in treating patients as persons, not diseases
	Question	Rating nurses' answers to patients' questions
	Name	Rating personal care
	Concern	Rating how comfortable it is to tell nurses of any concerns
Professional Practice		
	Auton_1	Autonomy of nursing practice
	Centra_1	Centrality of decision making
	Colab_1	Collaborative relationship established between nurses and physicians
Job Satisfaction		
	Satisf_1	Work satisfaction of nurses
	Team_1	Satisfaction with team work

3. Specification in Analytical Modeling

A measurement model for each of the latent constructs is formulated to be examined in two-level analysis. First, patient satisfaction is a common factor that can be measured jointly by nine indicators. Patient satisfaction is labeled as PTSAT at the nursing unit level and as PSAT at the hospital level. Second, nursing professional practice is a common factor shared by three indicators (auton_1, centra_1, and colab_1). At the nursing unit level, professional practice is labeled as PROFPRAC; at the hospital level, it is labeled as PROF. Third, nurses' job satisfaction is a common factor shared by two indicators of work satisfaction. JSATISF is a nursing-unit level variable, and JSAT is a hospital-level variable for job satisfaction.

A covariance structure model is postulated for the two-level analysis.
The nursing unit level/ first level analysis:
PTSAT = F (JSATISF, BEDS, EXPER_1, AVAIL_1) + error_1.
J SATISF= F (PROFPRAC, TECH_1, IRNPROP) + error_2.
The hospital level/second level analysis:
SAT = F (JSAT, BEDS, EXPER_1, IRNPROP) + error_3.
JSAT = F (PROF, TECH_1) + error_4.
The multilevel covariance model can be further specified if exogenous variables such as hospital characteristics or nursing unit characteristics are included as predictor variables of the latent variables.

4. Selection of the Intervention Study Design and Analysis

This is not an intervention study. The analysis focuses on the effects that nursing unit and hospital measures of professional practice and job satisfaction have on patient satisfaction at the aggregated level.

5. Confirmatory Analysis

The analytical method used to formulate and validate the measurement model is confirmatory factor analysis (CFA). The advantage of CFA over exploratory factor analysis is that the investigator can impose substantively motivated constraints. These constraints determine (a) which pairs of common factors are correlated, (b) which observed variables are affected by which common factors, (c) which observed variables are affected by a unique factor or measurement error, and (d) which pairs of unique factors are correlated. Statistical tests can be performed to determine if the sample data are consistent with the imposed constraints or to evaluate whether the data confirm the general measurement model.

Before examining the effects of professional practice and job satisfaction on patient satisfaction, each measurement model should be validated. For example, the measurement model of patient satisfaction is presented in Figure 73. This figure can be redrawn for both between- and within-level analyses with a common factor shared by the nine indicators.

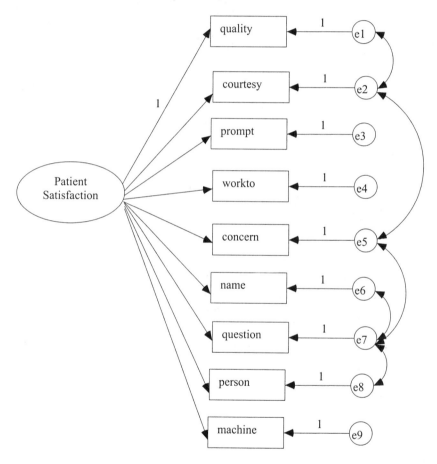

Figure 73. Measurement Model of Patient Satisfaction at the Nursing Unit Level

In this model, correlated error terms are assumed and validated by the two-level confirmatory factor analysis. Unlike conventional single-level analyses, where independence of observation is assumed for all N observations, in two-level analysis, independence is assumed only over the groups (Muthén, 1994). The variance components of the between-factor dimension (between hospitals) and the within-factor dimension (within nursing units) are independent. We cannot simply use the sample covariance matrix for estimating the population parameters, because nursing units are nested in hospitals. We must develop two separate matrices: (a) S_B refers to the covariance matrix of the disaggregated group (hospital) means, and (b) S_{PW} refers to the pooled within-group sample covariance matrix for nursing units. The procedures for performing CFA with Mplus are as follows:

1. Compute intraclass correlation coefficients for the observed indicators and determine whether the unit observations are independent for the hospitals studied. A coefficient greater than 0.10 suggests that two-level analysis is needed.
2. Compute a covariance matrix for the observed indicators for nursing units clustered in the hospitals, using SPSS program or Mplus.
3. Compute a pooled within-group covariance matrix, using Mplus.
4. Evaluate the between-group and within-group covariance models separately, so that a better-fitted model can be identified.
5. Incorporate both between- and within-group models, to fit the measurement model.

Table 39 summarizes the results of CFA with two-level analysis, using nine indicators of patient satisfaction measured at the nursing unit (group) level. All of the nine indicators show statistically significant associations with the common factor (e.g., patient satisfaction), irrespective of the level of analysis. This finding suggests that the nine indicators at both hospital and nursing unit levels can effectively measure patient satisfaction. Furthermore, to construct a composite index of patient satisfaction with the nine indicators is a viable option.

Table 39. Parameter Estimates for the Two-Level Analysis of Nine Indicators of Patient Satisfaction

Level of Analysis/ Indicators	Estimate	Standard Error	Critical Ratio
Hospital Level: n=60			
Quality	1.000		
Courtesy	0.957	0.092	10.411
Prompt	1.019	0.105	9.723*
Workto	0.836	0.090	9.329*
Concern	0.451	0.080	5.633*
Name	0.303	0.113	2.677*
Person	0.419	0.102	4.121*
Machine	0.748	0.082	9.124*
Nursing Unit Level: n=120			
Quality	1.00		
Courtesy	1.40	0.22	6.28*
Prompt	1.11	0.24	4.57*
Workto	0.93	0.20	4.61*
Concern	0.95	0.26	3.63*
Name	0.81	0.34	2.35*
Person	1.29	0.35	3.66*
Machine	1.27	0.29	4.38*
Tests of Model Fit: Chi-square=32.80. D.F.= 27. P value= 0.20. GFI=0.89. AGF=0.81. RMSEA=0.06.			

* Statistically significant at 0.05 or lower level

6. Establishment of Causality

The causal inference can be categorized into three levels: 1) individual effect of personal characteristics, 2) aggregate or ecological effect of personal characteristics, and 3) contextual effect of organizational or environmental factors. Using the previous example (nursing administrative and patient care outcomes), the first-level analysis will deal with the effects of nursing unit characteristics at the nursing unit level, and the second-level analysis will examine the ecological effects of aggregated characteristics of nursing units, and the contextual effects of hospital characteristics such as size, organizational change, and HMO penetration in a hospital area.

If the effects of nursing unit characteristics on patient satisfaction at both levels of analysis are confirmed, causal inference can reflect the importance of specific unit characteristics such as professional practice and nurses' job satisfaction, while the effects of other factors are simultaneously controlled. If the effects on units' patient satisfaction are confirmed only at the hospital level, not at the unit level, one cannot conclude that specific unit characteristics are true causal factors of patient satisfaction.

CONCLUSION

It is important to point out some of the differences between a regression-based hierarchical linear model (HLM) and a covariance structure based multilevel model (MLM). For example, in HLM, software is available to support 3-level models, while in MLM it is only limited to two-level models. In HLM, one can explicitly model cross-level interactions, while in MLM one cannot. Slopes as outcomes are not applicable in MLM. MLM identifies sources of variance in endogenous variables in terms of measurement errors at both the within and between levels, and structural sources of variation at both within and between levels. In addition, MLM can incorporate latent variables in the analysis, while HLM cannot.

The hierarchical structure of personal and organizational level data must be carefully considered in multivariate statistical analysis. Otherwise, the effects of personal and organizational characteristics on patient outcomes or organizational performance measures may be inappropriately estimated (Mark, Salyer, and Wan, 2001). Our example illustrates the complexity of designing and analyzing multilevel data in health services management and research. The analytical framework of the context-design-performance-outcome model specified earlier should be validated with multilevel data. This will permit proper generalizations of the effects of personal and organizational factors on patient care outcomes.

REFERENCES

Bryk, A.S., Raudenbush, S.W. (1992). *Hierarchical Linear Models*. Newbury Park, CA: Sage Publications.

Dimatteo, M.R., DiNicola, D.D. (1983). *Achieving Patient Compliance: The Psychology of the Medical Practitioner's Role*. NY: Pergamon, Press.

Donabedian, A. (1980). *Explorations in Quality Assessment and Monitoring. Vol 1. The Definition of Quality and Approaches to its Assessment*. Ann Arbor, MI: Health Administration Press.

Eriksen, L.R. (1995). Patient satisfaction with nursing care: Concept clarification. *Journal of Nursing Measurement* 3(1): 59-76.

Heck, R.H., Thomas, S.L. (2000). *An Introduction to Multilevel Modeling Techniques*. Mahwah, NJ: Lawrence Erlbaum Associates.

Little, T.D., Schnabel, K.U., Baumert, J. (2000). *Modeling Longitudinal and Multilevel Data*. Mahwah, NJ: Lawrence Erlbaum Associates.

Mark, B.A., Salyer, J., Wan, T.T.H. (2001). Professional nurse practice: impact on organizational and patient outcomes. *Medical Care (under revision)*.

Morgenstern, H. (1998). Ecologic studies. In Kenneth J. Rothman and Sander Greenland (eds.), Modern *Epidemiology*. New York: Lippincott Williams and Wilkins.

Muthén, L.K., Muthén, B.O. (1998). *Mplus: User's Guide*. Los Angeles: Muthén and Muthén.

Muthen, B.O. (1994). Multilevel covariance structure analysis. *Sociological Research Methods and Research* 22(3): 376-399.

Rosenthal, G.E., Shannon, S.E. (1997). The use of patient perceptions in the evaluation of health care delivery systems. *Medical Care* 35(11): NS58-NS68.

Shelton, P.J. (2000). *Measuring and Improving Patient Satisfaction*. Gaithersburg, MD: Aspen Publishers.

Thomas, J.W., Penchansky, R. (1984). Relating satisfaction with access to utilization of services. *Medical Care* 22(6): 553-568.

Vuori, H. (1987). Patient satisfaction: An attribute or indicator of the quality of care? *Quality Review Bulletin* 13:106.

Wan, T.T.H. (1995). *Analysis and Evaluation of Health Care Systems: An Integrated Approach to Managerial Decision Making*. Baltimore, MD: Health Professions Press.

CHAPTER 12

GROWTH CURVE MODELING WITH LONGITUDINAL DATA

Researchers examining the performance of health care organization have not made serious attempts to examine how time-invariant and time-varying factors affect efficiency and financial viability, by using longitudinal analysis. Analyses of cross-sectional hospital performance data have shed little light on the explanatory power of the predictor variables. This chapter introduces a multivariate modeling strategy analyzing multi-wave data collected from a panel of health care organizations, particularly hospitals.

Linear growth curve modeling is useful for investigating the dynamic relationships between multiple causes and outcomes of health care organizations. Its flexible, multivariate framework is richer and more versatile than other analytical modeling methods for studying multiple growth processes of individual organizational units (McArdle and Epstein, 1987; Muthén, 1991; Muthén, 1997). Wan et al. (2001) studied the relationship between integration strategies and the performance of integrated health care networks. They found that little variation in measures of productivity, finance and image among the top 100 health care networks could be attributed to integration strategies. Their study is constrained by the use of two-wave data. However, if three or more waves of longitudinal data on network performance are available, a growth curve model would be a fruitful approach to understanding how one or more performance variables or processes develop over time.

INTRODUCTION TO GROWTH CURVE MODELING

In growth curve modeling, random coefficients are used to capture individual differences in development and performance. In a structural equation modeling framework, the random coefficients are conceptualized as continuous latent variables, that is, two types of growth factors: the initial level and the growth trend. The first growth factor, which is one random coefficient (η_{0i}), describes the intercept of the growth curve, which refers to the initial level of the growth curve. The second type of growth factor, which may consist of one or more (slope) coefficients, describes the shape of the growth curve (η_{1i}). The slope (S) can be treated as a quadratic element (S^2) of the growth curve, to show the trajectory of the change in performance or outcome. For a given performance or outcome variable (Y_{it}), an equation can be written as follows: $Y_{it} = \eta_{0i} + \eta_{1i} X_t + \varepsilon_{it}$.

The intercept (η_{0i}) is a linear combination of three elements: $\alpha_0 + \Upsilon_0 W_1 + \zeta_{01}$, where, α_0 is the mean, Υ_0 is the effect of an exogenous variable (W_i) on η_{0i} at the initial level, and ζ_{01} is a residual term. The slope (η_{1i}) also has three elements: $\alpha_1 + \Upsilon_1 W_i + \zeta_{1i}$, where α_1 is the mean, Υ_1 is the effect of an exogenous variable (W_i) on η_{1i}, and ζ_{1i} is a residual term for this equation.

A growth curve model can be viewed as a restricted factor analytic model consisting of nonzero factor means (MacCallum and Kim, 2000). This model can be fit to means and covariances of performance or outcome variables by using Mplus (Muthén and Muthén, 1998) or LISREL 8 (Jöreskog and Sörbom, 1994).

Mplus can estimate the following growth models: linear, quadratic, growth models with free time scores, parallel process growth models, sequential process growth models, multiple indicator growth models, multiple groups, growth models with regressions among random coefficients, and growth models with interventions.

Growth models can have both time-invariant and time-varying covariates. For example, hospital efficiency measured at four time points can be predicted by time-invariant variables such as bed size, ownership, rural/urban location, and system affiliation, and also by time-varying or time-specific variables such as nurse skill mix (RN ratio) at each time point studied. Mplus also can estimate growth models for continuous observed variables; categorical observed variables; and continuous latent variables with continuous, categorical (binary or ordered polytomous) variables; or for a combination of continuous and categorical factor indicators. When observed variables or factor indicators are all continuous, Muthén and Muthén (1998) note that Mplus has five estimator choices: maximum likelihood (ML); maximum likelihood with robust standard errors and chi-square (MLM, MLMV); generalized least squares (GLS); and weighted least squares (WLS), also referred to as ADF. When at least one observed variable or factor estimator is categorical, Mplus has four estimator choices: weighted least squares (WLS), robust weighted least squares (WLSM, WLSMV), and unweighted least squares (ULS). Detailed information on Mplus can be found in www.statmodel.com.

RELATED RESEARCH ON COMPLEX HELTH CAREORGANIZATIONS' PERFORMANCE

Scott (1993) and Luke and Begun (1988) noted that health care managers choose services and design systems to maximize effectiveness and efficiency. They also point out that an organization's strategy should be consistent with both external environmental demands and the organization's core capabilities and competencies. What remains less clear, however, is how the organization's choice of strategies affects its performance.

Lin and Wan (1999) classified the strategic directions of an IHN into the three categories suggested by Hofer and Venkatraman (1978) in their analysis of organizational strategies: 1) an IHN's corporate strategy, enlarging the network size; 2) its business strategy, venturing into non-hospital services; and 3) its functional strategy, integrating information systems and financial arrangements for cooperative purchases. Still largely missing from functional integration is clinical integration, which is a prerequisite of a fully integrated delivery system. Efforts have been made to deliver coordinated care through case management and disease management. However, a truly integrated delivery system should optimize its functions by achieving structural (administrative and managerial), informatic, financial and clinical integration (Wan, 1995).

Gillies and associates (1993) conducted a Health System Integration Study of twelve integrated delivery systems. Intensive site visit interviews were conducted in each system. The authors found a moderate level of integration overall, particularly in culture, financial planning, and strategic planning. The levels of physician-system integration and clinical integration, however, were low. Moreover, information system and nonclinical support activities were found to be the least integrated. From Spearman-rank correlation analysis, significant associations were reported among structural integration, physician-system integration, and clinical integration. The study also found a positive relationship between perceived integration and perceived effectiveness. Shortell and his associates (1993) identified eight major barriers to greater levels of integration: (1) failure to understand the new core business, (2) inability to overcome the hospital paradigm, (3) inability to convince the "cash cow" to accept system strategy, (4) inability of the board to understand the new health care environment, (5) ambiguous roles and responsibilities, (6) inability to "manage" managed care, (7) inability to execute the strategy, and (8) lack of strategic alignment.

In a study of horizontally integrated hospitals, McCue, Clement and Luke (1999) examined the relationship between the type of strategic hospital alliance (SHA) and financial performance, with market, environmental, and operational factors simultaneously considered. They found that SHAs did not reap much financial benefit from alliances. Findings about urban hospitals in Clement et al. (1997) suggest that membership in a strategic hospital alliance contributes to the individual hospital's revenues, but does not promote individual hospital efficiency.

Bazzoli and associates (2000) examined the relationship between organizational structure and financial performance in 1,047 health network hospitals and in 1,112 health system hospitals. They found that hospitals in health systems with single ownerships generally perform better financially than hospitals in contractually based health networks do. Among the health network hospitals studied, those belonging to highly centralized networks did better financially than those in more decentralized networks did. However, of the three types studied, the health system hospitals in moderately centralized systems performed best – i.e. better than those in highly centralized systems. Hospitals in networks or systems with little differentiation or centralization had the poorest financial performance.

Wan and associates (2001) analyzed the top 100 integrated health care networks (IHNs) ranked by the SMG study in 1998-1999, to identify the factors influencing INHs' performance and images. Using the Mplus program in an analysis of the two-wave data, they found that the technical efficiency score was inversely related to average length of stay and information integration. Other organization-based predictor variables had no statistical relationship with efficiency. Profit margin was related to two predictor variables, tax status and forward integration. For-profit IHNs yielded more profit margins than did non-profit networks. IHNs with forward integration, i.e. that ventured into multiple non-hospital-based services such as subacute care and long-term care, tended to have poorer profit margins than did those with no forward integration. An IHN's reputation or image as shown by ranking on the lists of the top 100 IHNs in both 1998 and 1999 was influenced by network size (the total number of facilities structured under one unified IHN), number of high-tech services offered, and implementation of coordinated care through case management and disease management programs. Bed size was inversely related to the IHN's reputation. The accuracy rate of predicting an IHN's reputation or image was 81%. The overall model fit was reasonable, with chi-square value of 42.754 (p <0.000).

In performing a longitudinal analysis (1980-1994) of the distinction between for-profit and nonprofit hospitals, Porter (2001) tested the effects that the hospital's legal charter or mission, regulatory environment, external environment, and internal factors serving as mediators had on cost efficiency and community service outcomes. Using growth curve modeling, Porter systematically demonstrated that nonprofit and for-profit hospitals were converging in terms of efficiency outcomes, and that nonprofit hospitals were simultaneously pursuing efficiency and community services.

In summary, growth modeling of longitudinal data offers a unique opportunity for researchers analyzing the changing hospital industry. This analytical approach enables us to examine multiple causes and multiple outcomes simultaneously over time. The following section illustrates the application of growth curve modeling.

APPLICATION

Data and Methods

Following the line of inquiry about the determinants of organizational performance, growth curve modeling is used to examine factors influencing hospital occupancy rates, an indicator of hospital performance. Data were obtained from the annual hospital surveys of the American Hospital Association for 1996-1999. After merging the data from four years, a total sample of 1092 hospitals is used to demonstrate growth modeling exercises.

The predictor variables of the occupancy rate were selected according to the availability of data from the American Hospital Association (AHA) files and were categorized into structural and operational characteristics. The hospital characteristics identified in 1996 were treated as the time-invariant variables. They are: 1) system affiliation status (SYS: affiliated hospital coded 1 and not- affiliated hospital coded 0); 2) HMO contract (HMO: with a contract coded 1 and without a contract coded 0); 3) metropolitan size (MSIZE: ranging from a rural area coded 0 to the largest metropolitan statistical area coded 6; 4) teaching status (TECH: teaching hospital coded 1 and non-teaching hospital coded 0); and 5) ownership (OWNER: for-profit hospital coded 1 and nonprofit hospital coded 0). The nurse skill mix, RN-nurse ratio (RN), is treated as an operational indicator that varies in each study year, a time-varying predictor variable of the occupancy rate (OCC).

A brief description of each of the postulated models is presented, along with an empirical analysis of multi-wave panel data for acute care hospitals in the United States. Each longitudinal analysis examines the patterns or trajectories of the changes among individual hospitals during four years (1996-1999).

Application of Growth Modeling

1. A Linear Growth Model of Four Occupancy Rates

Figure 74 presents a single-indicator growth model. Occupancy rates are measured at four occasions. The model is intended to account for the means, variances, and covariances of four repeated measures of performance. Two factors are specified, an intercept (I) and a slope (S) factor; the two factors are assumed to be correlated. The loadings for the intercept factor are all fixed to 1.0, indicating that this factor has a fixed and equal influence on all measures of hospital occupancy. The loadings for the slope factor are fixed at the known values of the time measure (in this case, the year indicators 0, 1, 2, and 3 for the period from 1996 to 1999). This pattern of loadings implies that the slope factor has an increasing effect on the successive measures of hospital occupancy. The Mplus program for this model is presented in Table 40.

Table 40. Mplus Program for Model 1

TITLE: A LINEAR GROWTH MODEL FOR HOSPITAL OCCUPANCY RATES
       (1996-1999): GROWTH_1
DATA: FILE IS "TEST.TXT";
VARIABLE: NAMES ARE HOSPID, BED, SYS, HMO, MSIZE,
       TEACH, OWNER, OCC96, LOS96, COST96, RN96,
       OCC97, LOS97, COST97, RN97, OCC98, LOS98, COST98, RN98, OCC99,
       LOS99, COST99, RN99;
       USEVARIABLES ARE OCC96, OCC97, OCC98, OCC99;
       MISSING IS .;
ANALYSIS: TYPE=MEANSTRUCTURE;
MODEL:
       I BY OCC96-OCC99@1;
       S BY OCC96@0, OCC97@1, OCC98@2, OCC99@3;
       [OCC96-OCC99@0, I,S];
OUTPUT: SAMP, STANDARDIZED, MOD(5);

This model has no predictor variables. Here we are interested in examining the goodness of fit for the growth model and also looking at the strength of associations among the growth parameters. This model has a chi-square value of 18.005 with 5 degrees of freedom (p=0.003, N=1093). The root mean squared error of approximation (RMESA=0.049, p=0.485) shows a good fit of the model. The linear-rate means in 1996 appear to be significantly different from 0 (critical value of 55 at 0.05 level), whereas the means for the slope are not statistically significantly different (critical value of 0.762). The linear growth variances for both intercept and slope are statistically significantly different from 0 (critical values of 22.138 and 10.499, respectively. A negative correlation (-0.102) exists between the two growth factors (I and S) with a critical value of-2.343. This implies that the hospitals having higher occupancy rates in 1996 tended to grow more slowly later years than did those having lower occupancy rates at the initial measurement occasion. Finally, the two growth factors are adequately measured by the four occupancy rates (1996-1999), with R-square values of 0.938, 0.901, 0.933, and 0.954 for the respective years.

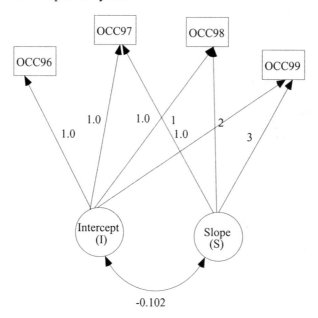

Figure 74. A Linear Growth Curve Model of Hospital Occupancy Rates (1996-1999): Model 1

2. A Linear Growth Model with Time –Invariant and Time-Varying Predictor Variables

Figure 75 presents a linear growth curve model of hospital occupancy rates with predicator variables, including both time-invariant and time-varying variables. Previous assumptions for the two growth factors are still valid in this model. By including five predictor variables (system affiliation, HMO contract, metropolitan population size, teaching status, and ownership) with direct causal paths with the two growth factors, we postulated that the organizational and contextual characteristics of hospitals measured in 1996 directly affected the two growth factors. In addition, a correlation between the two residual terms for the intercept and slope was tested. A time-specific or time-varying predictor variable (RN ratio) was introduced to affect the occupancy rate. The residual terms of the occupancy rates were independent, not correlated. Table 41 shows the Mplus program for this model.

Table 41. A Mplus Program for Model 2

```
TITLE: A LINEAR GROWTH MODEL FOR HOSPITAL OCCUPANCY RATES
       (1996-1999) WITH TIME-INVARIANT & TIME-VARYING PREDICTORS:
       GROWTH_2
DATA: FILE IS "TEST.TXT";
VARIABLE:  NAMES ARE HOSPID, BED, SYS, HMO, MSIZE,
       TEACH, OWNER, OCC96, LOS96, COST96, RN96,
       OCC97, LOS97, COST97, RN97, OCC98, LOS98, COST98,
       RN98, OCC99, LOS99, COST99, RN99;
       USEVARIABLES ARE OCC96, OCC97, OCC98, OCC99, SYS, HMO,
       MSIZE, TEACH, OWNER, RN96, RN97, RN98, RN99;
       MISSING IS .;
ANALYSIS:  TYPE=MEANSTRUCTURE;
MODEL:
       I BY OCC96-OCC99@1;
       S BY OCC96@0, OCC97@1, OCC98@2, OCC99@3;
       [OCC96-OCC99@0, I,S];
       I S ON SYS, HMO, MSIZE, TEACH, OWNER;
       OCC96 ON RN96;
       OCC97 ON RN97;
       OCC98 ON RN98;
       OCC99 ON RN99;
OUTPUT:    SAMP, STANDARDIZED, MOD(5);
```

The model has a chi-square value of 97, with 27 degrees of freedom (p=0.000, N=1004). The root mean squared error of approximation (RMESA=0.051, p=0.431) indicates that the model is well fit. The intercept (I) is positively influenced by three of the five predictor variables: metropolitan population size, teaching status, and system affiliation, in order of their relative importance (Figure 75). In the initial year (1996), the occupancy rates were higher in hospitals located in the larger metropolitan population areas (by 3.2%), in teaching facilities (by 16.5%), and in those affiliated with health systems (by 5.4%) than they were in their respective counterparts. These characteristics exerted no direct influence on the slope growth factors, as they were not statistically significant at the 0.05 level. The time-varying predictor "RN ratio" appears to influence the occupancy rates of

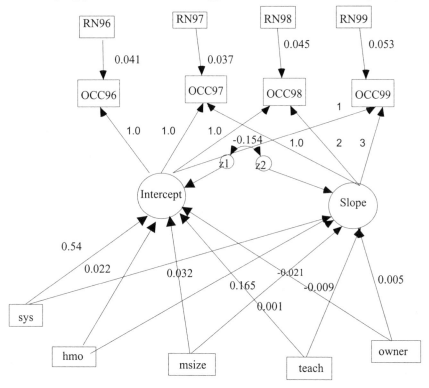

Figure 75. A Linear Growth Curve Model of Hospital Occupancy Rates (1996-1999) with Time-Invariant and Time Varying Predictors: Model 2

1997-1999, with statistical significance at the 0.05 level. A negative correlation (-0.102) exists between the two residual terms of the growth factors (I and S), having a critical value of -3.940. This implies that a common source of variations in the two growth factors may be accounted for by an unidentified predictor variable. Finally, the variance in occupancy rates at the initial time can be explained adequately by the five predictor variables (R^2=0.314). However, only a negligible amount of variance in the growth trend of occupancy rates (R^2=0.028) was accounted for by the predictor variables.

3. A Parallel Process Growth Model for Two Continuous Variables

The parallel process growth model addresses the question: Are the growth patterns of occupancy rates and RN ratios similar over the time span of four years? To answer this question, we have to test the concomitant presence of the two parallel growth factors (intercepts and slopes) for the two continuous dependent variables, occupancy rate and RN ratio.

This model, Figure 76, has the following assumptions: 1) the level growth factor (I2) of RN ratios affects the level growth factor (I1) of occupancy rates; 2) the level growth factor (I2) of RN ratios directly influences the trend growth rate (S1) of occupancy rates; 3) the level growth factor (I1) of occupancy rates directly influences the trend growth rate (S2) of RN ratios; 4) the two residual terms of the intercepts are correlated; 5) six time-invariant variables or covariates directly affect the intercepts (I1 & I2) and slopes (S1 & S2) of the outcome variables; 6) residuals of the first growth process's first two outcome measures (OCC96 with OCC97) are correlated; 7) residuals of the trend growth factors are also correlated by default, given that the trend growth factors do not influence other latent variables; and 8) no correlated error terms are observed between occupancy rates and RN ratios. The Mplus program for this model is shown in Table 42.

This model has a chi-square value of 53.103, with 44 degrees of freedom (p=0.163, N=1092). The root mean squared error of approximation (RMESA) is 0.014, with p value of 1.000. The level growth factors (I1 and I2) of the two outcome variables (occupancy rate and RN ratio) are positively correlated (0.245) at the 0.05 level (critical value of 7.662). The trend growth factor (S1) of occupancy rates is not related to the level growth factor (I2) of RN ratios. The trend growth factor (S2) of RN ratios is not related to the level

Table 42. Mplus Program for Model 3

```
TITLE: A PARALLEL PROCESS GROWTH MODEL FOR HOSPITAL OCCUPANCY
       RATES  &  RN  RATIOS  (1996-1999)  WITH  TIME-INVARIANT
       PREDICTORS: GROWTH_3
DATA: FILE IS "TEST.TXT";
VARIABLE:  NAMES ARE HOSPID, BED, SYS, HMO, MSIZE,
       TEACH, OWNER, OCC96, LOS96, COST96, RN96,
       OCC97, LOS97, COST97, RN97, OCC98, LOS98, COST98,
       RN98, OCC99, LOS99, COST99, RN99;
       USEVARIABLES ARE OCC96, OCC97, OCC98, OCC99, SYS, HMO,
       MSIZE, TEACH, OWNER, RN96, RN97, RN98, RN99;
       MISSING IS .;
ANALYSIS:  TYPE=MEANSTRUCTURE;
MODEL:
       I1 BY OCC96-OCC99@1;
       S1 BY OCC96@0, OCC97@1, OCC98@2, OCC99@3;
       I2 BY RN96-RN99@1;
       S2 BY RN96@0, RN98@1, RN99@2, RN99@3;
       [OCC96-OCC99@0, I1, S1, I2, S2];
       I1 I2 S1 S2 ON SYS, HMO, MSIZE, TEACH, OWNER;
       S1 ON I2;
       S2 ON I1;
       I1 ON I2;
       OCC96 WITH OCC97;
       RN96 WITH RN98;
       RN97 WITH RN98;
       RN96 WITH RN99;
OUTPUT:    SAMP, STANDARDIZED, MOD(5);
```

growth factor (I1) of occupancy rates. The statistically significant predictor variables for the level growth factor of occupancy rates are: 1) bed size (7.613), 2) metropolitan population size (7.194), 3) teaching status (3.565), and 4) system affiliation (2.714). The six predictors combined account for 42.7 percent of the total variance in the level growth factor of occupancy rates. The four predictors that are statistically significant in accounting for the variation in the level growth factor (I2) of RN ratios are: 1) metropolitan population size (6.986), 2) HMO contract (4.528), 3) ownership (-4.115), and 4) system affiliation (3.596).   The six predictors account for 18.3 percent of the variation in the level growth factor of RN ratios. These same predictors account for a negligible amount of the variations in the trend growth factors ($R^2=6.3\%$).   The four correlated residuals of the outcome variables are statistically significant (OCC96 with metropolitan population size (6.986), 2) HMO contract (4.528), 3) ownership (-4.115), and 4) system affiliation (3.596). The six predictors account for 18.3 percent of the variation in the level growth factor of RN ratios. These same predictors account for a negligible amount of the variations in the trend growth factors (R^2=6.3%).

The four correlated residuals of the outcome variables are statistically significant (OCC96 with OCC97: 4.587; RN97 with RN98: 3.675; RN96 with RN98: -3.225; RN96 with RN99: -4.894. The residuals of the two-trend growth factors are not statistically significant.

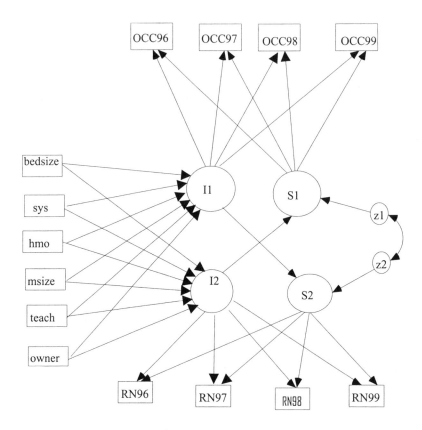

Figure 76. A Parallel Process Growth Model for Two Continuous Outcome Variables with Time-Invariant Predictors (1996-1999): Model 3

CONCLUSION

The Balanced Budget Act (BBA) mandated budget savings of $160 billion over five years. The comprehensive spending cuts enacted on August 5, 1997 have caused rapid deflation in Medicare pricing, with an estimated budgetary savings of 30% in Medicare for the period of 1998-2002. Consequently, the BBA may have directly affected hospital occupancy rates and RN ratios, reducing both.

The linear growth curve models were developed to analyze two continuous outcome variables. The key findings are summarized as follows: 1) the level growth factors of hospital occupancy rates and RN ratios are positively related; 2) hospitals with large bed size, with teaching and system affiliation, and with locations in large metropolitan areas appeared to have higher initial (level) growth factors for occupancy rates; 3) higher RN ratios are found in hospitals that are affiliated with health care systems or networks, have HMO contracts, are located in large metropolitan areas, or are identified as nonprofit facilities; 4) hospital characteristics and contextual variables do not explain the trend growth factors of occupancy rates and RN ratios; and 5) the parallel process growth model for occupancy rates and RN ratios is supported by the data analyzed.

Growth modeling of four waves of hospital data shows that the level growth factors (intercepts) of both outcome variables are the dominant aspect of the growth pattern, and that the trend growth rates (slopes) are not very important in explaining the parallel process growth patterns of occupancy rates and RN ratios in the period of four years (1996-1999). These findings suggest that, in fact, the Balanced Budget Act may have had little effect on the change rates of hospital occupancy and RN ratios. Although the RN ratio is positively related to the occupancy rate, the trend growth rate of occupancy rates is not related to the level growth factor of RN ratios, and the trend growth rate of RN ratios is not related to the level growth factor of occupancy rates. This indicates that the growth trends or change rates of these two outcome variables are not well explained by the model. Future studies on changes in hospital operations should include the supply and demand variables as well as market characteristics such as managed care penetration and population health measures.

The growth modeling analysis of longitudinal data is a very flexible and versatile statistical tool. This multivariate approach to repeated measures of hospital performance could be a standard, confirmatory analytical procedure for analyzing continuous or discrete outcome variables. Mplus programs, particularly growth modeling with time-invariant and time-varying predictor variables, will allow researchers to test a variety of theoretical models in organizational research.

REFERENCES

Bazzoli, B.J., Chan, B., Shortell, S., D'Aunno, T. (2000). The financial performance of hospitals belonging to health networks and systems. *Inquiry* 37 (3): 234-252.

Bellandi, D. (March 29, 1999). Ranking the networks. *Modern Health care* 60-64.

Clement, J.P, McCue, M.J., Luke, R.D., Ozcan, Y. et al. (1997). Strategic hospital alliances: impact on financial performance. *Health Affairs* 16:193-203.

Coddington, D.C., Moore, K.D., Fischer, E.A. (1994). Costs and benefits of integrated health care systems. *Health Care Financial Management* 48 (3): 20-29.

Conrad, D.A. (1992). Vertical Integration. In W. J. Duncan, P. Ginter, and L. Swayne (eds.), *Strategic Issues in Health Care: Point and Counterpoint*. Boston: PWS Kent.

Conrad, D.A., Shortell, S.M. (1996). Integrated health systems: Promise and performance. *Frontiers of Health Service Management* 13 (1): 3-40.

Gillies, R.R., Shortell, S.M., Anderson, D.A., Mitchell, J.B., Morgan, K.L. (1993). Conceptualizing and measuring integration: findings from the health systems integration study. *Hospital and Health Services Administration* 38 (4): 467-489.

Hofer, C.W., Venkatraman, N. (1978). *Strategy Formation: Analytical Concepts*. St. Paul, MN: West.

Hult, G.T., Lukas, B.A., Hult, A.M. (1996). The health care learning organization. *Journal of Hospital Marketing* 10 (2): 85-99.

Jöreskog, K.G., Sörbom, D. (1994). *LISREL 8 User's Guide*. Chicago: Scientific Software.

Lin, Y.J., Wan, T.T.H. (2001). Effect of organizational and environmental factors on service differentiation strategy of integrated health care networks. *Health Services Management Research* 14: 18-26.

Lin, Y.J., Wan, T.T.H. (1999). Analysis of integrated health care networks' performance: A contingency-strategic management perspective. *Journal of Medical Systems* 23(6): 477-495.

Luke, R.D., Begun, J. W. (1988). The management of strategy. In Shortell, S. M. and Kaluzny, A.D. (eds.), *Health Care Management: A Text in Organizational Theory and Behavior*. New York: John Wiley and Sons, Inc.

McCue, M.J., Clement, J.P., Luke, R. (1999). Strategic hospital alliances: Do the type and market structure of strategic hospital alliances matter? *Medical Care* 37(10): 1013-1022.

McArdle, J.J., Epstein, D. (1987). Latent growth curves within developmental structural equation models. *Child Development* 58: 110-133.

Muthén, B.O. (1991). Analysis of longitudinal data using latent variable models with varying parameters. In Collins, L. and Horn, J. (eds.), *Best Methods for the Analysis of Change. Recent Advances, Unanswered Questions, Future Directions*. Washington, D.C.: American Psychological Association, pp. 1-17.

Muthén, B.O. (1997). Latent variable modeling with longitudinal and multilevel data. In Raftery, A. (ed.), *Sociological Methodology*. Boston: Blackwell Publishers, pp. 43-480.

Muthén, L.K., Muthén, B.O. (1998). *Mplus User's Guide: The Comprehensive Modeling Program for Applied Researchers*. Los Angeles: Muthén & Muthén.

Porter, S J. (2001). A longitudinal analysis of the distinction between for-profit and non-for-profit hospitals in America. *Journal of Health and Social Behavior* 42 (1): 17-44.

Scott, W.R. (1993). The organization of medical care services: toward an integrated theoretical model. *Medical Care Review* 50(3): 271-302.

Sexton, T.R. (1986). *The Methodology of Data Envelopment Analysis*. San Francisco: Jossey-Bass.

Shortell, S.M., Gillies, R.R., Anderson, D.A., Mitchell, J. B., Morgan, K. L. (1993) Creating organized delivery systems: the barriers and facilitators. *Hospital and Health Services Administration* 38 (4): 447-466.

Wan, T.T.H. (1995). *Analysis and Evaluation of Health Care Systems: An Integrated Managerial Decision Making Approach*. Baltimore, MD: Health Professions Press.

Wan, T.T.H., Ma, A., Lin, B.Y.J. (2001). Integration and the performance of health care networks: Do integration strategies enhance efficiency, profitability, and image? *International Journal of Integrated Care* 1 (3):1-10.

Zeller, T.L., Stanko, B.B., Cleverley, W. (1997). A new perspective on hospital financial ratio analysis. *Health Care Financial Management* 51 (11): 62-66.

CHAPTER 13

EPILOGUE

Previous chapters have illustrated how structural equation modeling can be applied to research in health services management. Examples and explanatory models were developed for understanding how contextual, structural, and design factors may influence organizational performance. The structural relationships among these variables were empirically examined by using LISREL, AMOS, and/or Mplus programs.

In this concluding chapter, several conceptual, methodological, and practical issues in the applications of structural equation modeling are noted. First, analytical models should be based on the parsimonious principle, to portray the reality of structural relationships among the study variables. Alternative or competing models should be developed and tested, to arrive at better model. One must realize that a confirmed model with non-experimental data should be treated only as an approximation of reality. The explanatory model can be used to design more rigorous experimental studies to verify and validate the theoretical model.

Second, using structural equation modeling for causal inference in organizational studies illuminates the complex structural relationships among multiple exogenous and endogenous variables. However, attention to refinement of measurement instruments or variables is essential, to reduce measurement errors (e.g., poorly formulated scales) and specification errors (e.g., inadequate operationalization or invalidity of the variables).

Structural equation modeling in health services management research is a heuristic rather than a predictive tool. New knowledge or understanding can be generated by the proper application of this useful analytical tool. Confirmatory analyses can generate the sort of consistent evidence that can guide the development of a support system for executive decision making (Figure 77). Such a system has an interface between two different approaches to knowledge management: 1) confirmatory analysis such as structural equation modeling to produce new knowledge, and 2) simulation and optimization methods such as the constraint-oriented reasoning methodology used in workflow analysis and managerial accounting.

Most importantly, researchers should build an interface between analytical modeling and operations research. It would be extremely helpful if researchers would design user-friendly, graphics-based presentations for practical use. Ultimately, health care executives or decision makers should come to rely on theoretically based, empirically validated hard evidence, rather than on tacit or experiential knowledge, when deciding on management strategy.

Health services management is an endlessly challenging field. The central contribution of structural equation modeling in knowledge management is to offer a means of gathering pertinent data, producing useful information, and creating vital knowledge that will enhance organizational performance.

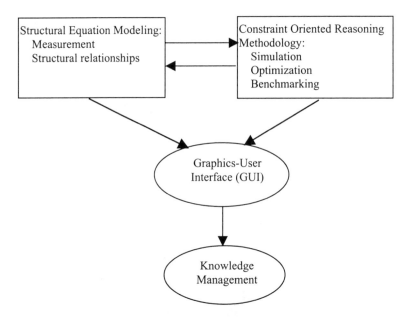

Figure 77. An Interface between Structural Equation Modeling and
Constraint-Oriented Reasoning Methodology, for
Health Care Knowledge Management and Research

INDEX

adjusted goodness of fit, 148, 149, 190, 191, 192, 194

adverse health events, 5, 76

AMOS, 73, 86, 108, 169, 187, 197, 230

analyzing program performance, 22

automatic interaction detector analysis, 40

autoregressive model, 118

Balanced Budget Act, 7, 224

bases of health care knowledge, 3

buffering effect, 162, 163

causal effect, 61, 119, 131, 133, 158

causal inference, 6, 11, 18, 19, 26, 27, 32, 37, 51, 203, 208, 209, 230

causal inquiry, 7, 14, 202

causal modeling, 7, 31

causality, xiv, 6, 14, 18, 19, 25, 26, 29, 30, 31, 32

cause-effect relationship, 18, 20, 25

classic experimental design, 21

coefficient of alienation, 61

complex experimental design, 21

conceptualization, 15

confirmatory analysis, 14, 203, 230

confirmatory factor analysis, 7, 55, 71, 73, 88, 93, 95, 101, 103, 106, 115, 136, 158, 169, 170, 188, 205, 207

constraint-oriented reasoning methodology, 230

construct validity, 93, 106, 184

context-design-performance framework, 172

correlated measurement errors, 109, 155, 191, 195

correlated residuals, 222, 223

covariance structure model, 7, 81, 154, 157, 159, 162, 163, 165, 167, 169, 184, 188, 190, 193, 204, 205

cross-sectional study, 19, 52, 167

data mining, 2, 5, 7, 8, 9, 27, 39, 42, 43

data warehousing, 4, 7, 9, 27, 39, 43

decomposition of causal effects, 133

deductive approach, 12

direct effect, 61, 67, 141, 158, 171

ecological fallacy, 28, 201

effectiveness, xiv, 13, 30, 35, 41, 143, 153, 214

effectiveness studies, 35

efficiency, 4, 7, 13, 36, 43, 84, 143, 144, 147, 148, 149, 150, 151, 152, 153, 165, 166, 167, 168, 169, 171, 173, 174, 185, 186, 197, 212, 213, 214, 215, 216, 228

endogenous variable, 18, 22, 23, 26, 27, 41, 60, 61, 62, 68, 81, 85, 118, 119, 127, 128, 134, 141, 143, 149, 154, 155, 167, 177, 178, 183, 186, 188, 209, 230

equipotency curves, 18

evidence-based disease management, 1

evidence-based health care, 1, 7, 41

evidence-based health care management, 7, 41

evidence-based management, 1, 35

evidence-based medicine, 1, 8

exogenous variable, 18, 21, 22, 41, 60, 61, 64, 81, 82, 83, 118, 119, 127, 128, 131, 138, 149, 151, 154, 158, 171, 176, 177, 187, 188, 193, 194, 205, 213

explicit knowledge, 3, 11

exploratory analysis, 40

factor analysis, 7, 31, 40, 54, 55, 57, 68, 71, 73, 80, 88, 93, 95, 97, 101, 103, 106, 109, 115, 136, 154, 158, 169, 170, 188, 205, 207

goodness of fit index, 149

goodness of fit statistics, 24, 40, 113

growth curve modeling, 41, 212, 216

growth factor, 212, 213, 218, 219, 220, 221, 222, 223, 224

growth modeling, 6, 7, 216, 225

health services management research, 6, 7, 36, 47, 93, 172, 230

health status, 10, 26, 33, 35, 45, 58, 74, 77, 93, 94, 106, 107, 108, 110, 113, 114, 115, 140, 159, 203

homoscendasticity, 76

hospital efficiency, 149, 167, 174, 197, 213, 215

hospital performance, 13, 35, 40, 83, 143, 144, 147, 151, 152, 153, 166, 167, 174, 212, 216, 226

hospital quality, 143, 197

hypothesis testing, 13, 23, 40, 128, 176

implicit knowledge, 3, 36

indirect effect, 61, 62, 67, 118

inductive approach, 11, 12

informatic integration, 4, 92, 166, 167, 169, 170, 171, 197

integrated care delivery systems, 7, 166

integrated health care networks, 153, 173, 212, 215, 227

interaction effect, 21, 47

intercept of the growth curve, 213

Joint Commission on Accreditation of, 11

knowledge management, 1, 2, 6, 7, 27, 29, 35, 40, 42, 166, 230, 231

lambda coefficient, 90, 91, 94

latent construct, 42, 76, 84, 89, 90, 93, 97, 101, 109, 110, 113, 114, 118, 147, 148, 162, 163, 170, 177, 179, 187, 202, 203, 205

latent variable, 6, 7, 9, 20, 23, 27, 41, 55, 73, 77, 78, 80, 81, 83, 84, 88, 94, 101, 108, 119, 136, 138, 139, 141, 154, 155, 156, 157, 162, 167, 169, 170, 187, 188, 204, 205, 209, 212, 213, 221, 227

LISREL, xi, 6, 20, 23, 30, 31, 40, 41, 44, 71, 72, 73, 77, 81, 82, 84, 86, 88, 94, 98, 100, 115, 118, 126, 127, 128, 129, 130, 131, 134, 139, 140, 144, 145, 147, 148, 151, 153, 176, 177, 188, 198, 213, 227, 230

longitudinal study, 140, 176

measurement errors, 62, 73, 78, 80, 83, 90, 91, 94, 101, 103, 104, 109, 110, 118, 136, 155, 190, 191, 195, 230

measurement model, 7, 9, 23, 41, 42, 71, 73, 74, 76, 77, 78, 82, 83, 84, 88, 89, 91, 92, 93, 94, 95, 97, 101, 103, 105, 114, 136, 141, 142, 145, 147, 148, 151, 158, 162, 163, 169, 170, 177, 178, 179, 184, 187, 188, 190, 191, 192, 193, 195, 204, 205, 206, 207

model, 6, 15, 16, 18, 20, 21, 22, 23, 24, 25, 26, 27, 28, 31, 32, 34, 39, 40, 41, 42, 44, 47, 52, 53, 55, 56, 57, 59, 60, 61, 62, 65, 66, 67, 68, 71, 72, 73, 76, 77, 78, 80, 81, 82, 83, 84, 88, 89, 91, 92, 93, 94, 95, 97, 98, 101, 103, 104, 105, 106, 109, 110, 113, 114, 118, 119, 120, 123, 127, 128, 130, 131, 132, 133, 134, 136, 137, 139, 140, 141, 142, 143, 144, 145, 147, 148, 149, 150, 151, 152, 154, 155, 156, 157, 158, 159, 162, 163, 164, 167, 169, 170, 176, 177, 178, 179, 181, 183, 184, 187, 188, 190, 191, 192, 193, 194, 195, 197, 198, 201, 202, 204, 205, 206, 207, 209, 212, 213, 216, 217, 218, 219, 220, 221, 224, 228, 230

Mplus, 73, 86, 203, 207, 210, 213, 215, 217, 219, 221, 222, 226, 227, 230

multicollinearity, 76, 93, 101

Multicollinearity, 76

multilevel analysis, 36, 202, 204

multilevel covariance modeling, 7

multi-level data, 28, 200

multiple group analysis, 7, 182, 183, 184, 188, 197

multiple logistic regression, 22, 52, 53

non-recursive model, 7, 118

Nursing Outcomes Study, 186

observability, 10, 11

occupancy rate, 7, 166, 216, 217, 218, 219, 220, 221, 222, 224

organizational research, 36, 91, 226

outcome indicators, 5, 19, 44, 115

outcome variables, 18, 22, 23, 40, 41, 71, 74, 83, 202, 213, 221, 222, 223, 224, 226

pain syndrome, 159

panel model, 118

path analysis, 6, 60, 61, 65, 68, 69, 118, 126, 127, 128, 131, 133, 134

patient care outcomes, 12, 37, 76, 83, 134, 139, 202, 208, 209

patient satisfaction, 7, 16, 30, 35, 122, 141, 178, 179, 184, 185, 186, 187, 188, 190, 193, 195, 196, 198, 203, 204, 205, 206, 207, 209

P-close, 82

physician productivity, 156

predictability, 77

predictability, 27

predictive validity, 77

professional practice, 201, 203, 204, 205, 206, 209

prospective study, 20

quality of care, xiv, 5, 6, 30, 76, 85, 152, 153, 166, 178, 179, 184, 187, 197, 203, 210

quantification, 14

quantification, 14, 16, 108, 167, 186, 204

quasi-experimental design, 20

recursive model, 7, 118, 137

regression model, 20, 22, 52, 118, 170

relational database, 1, 2, 37, 39

reliability, 17, 93, 101, 103, 106, 108, 113, 114, 115, 133, 184, 188, 203

residual path coefficient, 61

retrospective study, 20

RN ratios, 221, 222, 223, 224

robust standard error, 213

root mean squared error of approximation (RMSEA), 82

scientific inquiry, 10, 11, 12, 42

second-order factor, 32, 44, 105, 159

sentinel health events, 12

SF-36, 44, 105, 106, 107, 108, 113, 114, 115, 116, 159, 174
specification, 14, 15, 18, 19, 22, 37, 62, 80, 82, 84, 93, 97, 137, 139, 140, 141, 188, 230
structural equation modeling, 6, 7, 30, 31, 42, 71, 76, 77, 80, 197, 212, 230, 231
study design, 9, 19, 20, 26, 28, 52, 77, 182, 197
tacit knowledge, 3, 8

theoretical model, 6, 7, 15, 24, 27, 84, 105, 158, 183, 226, 228, 230
time-varying variable, 219
tractability, 10
validity, 17, 18, 62, 77, 93, 105, 106, 107, 108, 114, 115, 184, 188, 203
verifiability, 10, 18
weighted least square, 213
well being, 34, 101, 177